Nuclear Waste:
The Problem That Won't Go Away

Nicholas Lenssen

Worldwatch Paper 106
December 1991

Sections of this paper may be reproduced in magazines and newspapers with acknowledgment to the Worldwatch Institute. The views expressed are those of the author and do not necessarily represent those of the Worldwatch Institute and its directors, officers, or staff, or of funding organizations.

Printed on recycled paper

Table of Contents

Introduction 5

Permanent Hazard 8

Health and Radiation 15

They Call It Disposal...20

Technical Fixes?...27

The Politics of Nuclear Waste...34

Beyond Illusion..44

Notes 50

Introduction

In December 1942, humanity's relationship with nature changed for all time. Working in a secret underground military laboratory in Chicago, the emigré Italian physicist Enrico Fermi assembled enough uranium to cause a nuclear fission reaction. He split the atom, releasing the inherent energy that binds all matter together. Fermi's discovery almost immediately transformed warfare, eventually revolutionized medicine, and created hopes of electricity "too cheap to meter." But his experiment also produced something that will persist in an extremely hazardous form for hundreds of thousands of years—a small packet of radioactive waste materials.[1]

A half century later, scientists have yet to find a permanent and safe way to dispose of Fermi's radioactive waste, which lies buried under a foot of concrete and two feet of dirt on a hillside in Illinois. Nor have they found how to safely dispose of the 80,000-odd tons of irradiated fuel and hundreds of thousands of tons of other radioactive waste—contaminated hardware, filter sludges, and other dangerous detritus—accumulated since then by the commercial generation of electricity from nuclear power. Governments continue to promote the use of nuclear power without having any sure knowledge that a solution to this haunting problem is near, or indeed that the problem can be solved at all. The deadly residue of the nuclear age that Fermi inaugurated may be our civilization's longest-lasting legacy.[2]

Only the natural decay process, which takes hundreds of thousands or even millions of years, diminishes the radioactivity of nuclear waste. The significance of this crucial fact—that the health consequences, also, would last for millennia—was overlooked for decades after the discov-

I am grateful to Diane D'Arrigo, Helmut Hirsch, David Lowry, Lydia Popova, Scott Saleska, Mycle Schneider, and Barry D. Solomon, as well as to my colleagues at Worldwatch Institute, for reviewing drafts of this paper. Special thanks to Maureen and Nathan Lenssen for their invaluable support.

6 ery of radioactivity in the late nineteenth century. At first, no one even suspected that radioactivity was dangerous. As time passed, though, scientists found that radiation can have ravaging effects on human health, even at low levels of exposure. And radioactive materials, carried by wind and water, can spread quickly through the environment. Radioactive wastes from the U.S. Department of Energy's weapons facilities, including those at Hanford, Washington and Oak Ridge, Tennessee, have turned up in tumbleweeds, coyotes, and frogs. Chernobyl's explosion contaminated reindeer in Sweden and sheep in England, thousands of kilometers away. Atmospheric nuclear bomb testing created a radioactive "fallout" that spread around the globe.[3]

Though Fermi's crude experiment was part of the effort to design an atomic bomb, and bomb making has left a deadly legacy of its own, it is civilian reactors that have created 95 percent of the radioactivity emanating from U.S. nuclear wastes, and perhaps a similar share for the world as a whole. Moreover, civilian waste has been piling up faster than military waste. Among civilian uses, it is electric power that has become the Pandora's Box, while medical and other uses of radioactive materials have added little to the nuclear waste stream. The cumulative discharge of irradiated fuel from nuclear electric plants around the world is three times what it was in 1980 and twenty times what it was in 1970—and now awaits decisions on disposal in more than 25 countries.[4]

From the beginning of the nuclear age, governments were slow to deal with the resulting accumulation of waste, often promising fledgling commercial nuclear industries that they would one day assume responsibility for the waste—in effect leaving no one responsible, and bequeathing the problem to future generations. Working with radioactive waste was "not glamorous; there were no careers; it was messy, nobody got brownie points for caring about nuclear waste," according to Carroll Wilson, first general manager of the U.S. Atomic Energy Commission (AEC).[5]

Faced with the urgency of the Cold War and the perceived need to keep their publics fully in support of their military nuclear programs, atomic energy officials in Britain, France, the Soviet Union and the United States downplayed the problems of waste while aggressively promoting production of nuclear weapons. At the same time, governments encouraged—in some cases forced—the creation of civilian nuclear industries, and in the late sixties, civilian reactor construction programs took off. But the boom lasted little more than a decade. In the late seventies, a record of safety problems and accidents, questions about health effects, soaring costs, and eroding public confidence slowed reactor construction—and brought a new importance to the growing problem of nuclear waste.[6]

Most governments had decided by the sixties that burying nuclear waste deep in the earth is the best means of protecting the public. During the past decade, these efforts to get geologic burial underway have been marked by a growing sense of urgency—as the commercial nuclear industry found its waste piles mounting and popularity waning. Yet even today, "solutions" are hardly close at hand. Like mirages, safe and permanent methods of isolating radioactive materials seem to recede from reach as societies begin to examine them closely. It is now clear that geologic burial cannot guarantee that these materials will remain sealed off from the biosphere. The plans proposed so far have been found to be vulnerable—over time—to one kind of disruption or another, ranging from tectonic crushing or chemical bursting of containers to the corrosive and contamination-spreading action of groundwater. Nor have past efforts worked as planned; old burial and storage sites have proved leaky, and nuclear weapons facilities have become major public health hazards and the focus of multi-billion dollar cleanup campaigns.

Government efforts to quickly prove that geologic disposal is viable now face serious technical questions about the ability to guarantee that radioactive materials can be isolated for hundreds of centuries. Meanwhile, public confidence in the institutions responsible for handling nuclear waste, some of which are closely connected to the agencies that promote nuclear power, is so low that even the study of a location as a potential burial site brings people to the streets in protest.

Nuclear power now provides about 5 percent of the world's energy, a level it will not greatly exceed in the near future as construction work winds down on a few dozen remaining projects. Nevertheless, efforts are underway to revitalize construction programs. Government officials and nuclear industry executives, realizing that the industry is likely to remain stalled until the waste issue is dealt with, are pressing for a quick solution. Yet pressures to revive the stalled commercial nuclear industry could compromise efforts to resolve the waste problem. Just as the first-generation nuclear power plants were built without a full understanding of their environmental and social ramifications, a rushed job to bury wastes may turn out to be an irreversible mistake.[7]

Due to the intense political controversy that now seems to attend any proposal involving nuclear power, whether in Taiwan or Ukraine, true progress on the waste issue may only come about when a consensus on the future of the industry is achieved. A decision to phase out, or even forego, the use of nuclear power may be a bitter pill; but it would be easier to swallow before a country has developed a still larger dependence on nuclear energy.

Tragically, the nuclear waste problem remains without a politically or even technically feasible solution. And the volume and radioactivity keep piling up, at faster rates than ever before, as a growing number of nuclear power plants and related facilities are closed. The waste problem has emerged as the central nuclear issue of the next millennium—and beyond—and the one that may determine the final verdict on the nuclear age.

Permanent Hazard

The term "radioactive waste" covers everything from intensely hot irradiated (used) reactor fuel to mildly radioactive clothes worn by operators. Each type of waste contains its own unique blend of hundreds of distinct unstable atomic structures called radioisotopes. And each radioisotope has its own life span and potency for giving off alpha and

"Commercial nuclear power plants account
for 95 percent of the radioactivity
from all civilian and military sources combined."

beta particles and gamma rays, which cause harm to living tissue.
Radioisotope half-lives can vary from a fraction of a second to millions of
years. (The half-life is the amount of time it takes for 50 percent of the
original activity to decay; after 10 half-lives, one one-thousandth of the
original radioactivity—an amount that can still be dangerous—would
remain.) This means, for example, that the radioisotope plutonium-239,
a major constituent of irradiated fuel with a half-life of 24,400 years, is
dangerous for a quarter of a million years, or 12,000 human generations.
And as it decays it becomes uranium-235, its radioactive "daughter,"
which has a half-life of its own of 710,000 years.[8]

Perhaps the most dangerous radioactive waste of all, in its overall threat
to life on earth, is irradiated uranium fuel from commercial nuclear
power plants. It accounts for less than 1 percent of the volume of all
radioactive wastes in the United States but for 95 percent of the radioactivity from all civilian and military sources combined. The typical commercial nuclear reactor discharges about 30 metric tons of irradiated fuel
per year, with each ton producing nearly 180 million curies of radioactivity and generating 1.6 megawatts of heat. Since many of the radioisotopes in irradiated fuel decay quickly, its output of radioactivity falls to
693,000 curies per ton in a year's time. Even after 10,000 years, 470 curies
exist in each ton. (See Table 1.) (A curie measures the intensity of radiation at a fixed moment in time; as a reference point, there is some evidence that the Chernobyl accident released 50 million curies, measured
ten days later.)[9]

The world's 413 commercial nuclear reactors, producing about 5 percent
of the world's energy, created some 9,500 tons of irradiated fuel in 1990,
bringing the total accumulation of used fuel to 84,000 tons—twice as
much as in 1985. (See Figure 1.) The United States is "home" to a quarter
of this, with a radioactivity of over 20 billion curies. (See Table 2.) Within
eight years, the global figure could pass 190,000 tons. Total waste generation from the nuclear reactors now operating or under construction
worldwide will exceed 450,000 tons before the plants have all closed
down in the middle of the next century, forecasts the U.N.'s International
Atomic Energy Agency.[10]

9

Table 1: Radioactivity and Thermal Output Per Metric Ton of Irradiated Fuel from a Light-Water Reactor

Age of Fuel	Radioactivity	Thermal Output
(years)	(curies)	(watts)
At Discharge	177,242,000	1,595,375
1	693,000	12,509
10	405,600	1,268
100	41,960	299
1,000	1,752	55
10,000	470	14
100,000	56	1

Sources: Ronnie B. Lipschutz, *Radioactive Waste: Politics, Technology and Risk* (Cambridge, Mass: Ballinger Publishing Company, 1980); J.O. Blomeke et al., Oak Ridge National Laboratory, "Projections of Radioactive Wastes to be Generated by the U.S. Nuclear Power Industry," National Technical Information Service, Springfield, Va., February 1974.

Most of the existing irradiated fuel is stored in large pools of cooling water alongside nuclear reactors. These sites were originally designed to hold only a few years' worth of waste, and space constraints have led to packing the spent fuel closer together and using air-cooled vessels, including dry casks, to hold older, cooler irradiated fuel. Dry casks (containers that rely on passive cooling) are considered a safer way to prevent the fuel from overheating than water-based systems, which need electric pumps to circulate cooling water.[11]

Some countries, such as France and the United Kingdom, reprocess their irradiated fuel. Reprocessing was originally developed to extract the fission by-product plutonium for atomic bomb production. But this process involves chemical procedures that also remove the remaining uranium (only about 3 percent of the original uranium is fissionable). The recovered uranium can then be enriched, and again used as fuel. But while reprocessing "disposes" of irradiated fuel by putting it back into

Thousand tons

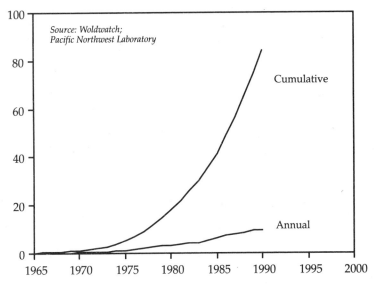

Source: Woldwatch;
Pacific Northwest Laboratory

Figure 1. World Generation of Irradiated Fuel from Commercial Nuclear Plants, 1965–1990

the power production process, it also leaves behind the most potent elements of it: the radioisotopes created by splitting uranium atoms—including cesium, iodine, strontium, and technetium, all of which constitute dangerous high-level waste. The remnants also include products of the intense neutron bombardment of nuclear fission—elements such as americium and neptunium, known as transuranics because they are heavier than uranium. Many of these isotopes have extremely long half-lives. Altogether, 97 percent of the radioactivity of used reactor fuel is still considered waste even after reprocessing. Thus, while reprocessing appears to reduce some waste by reclassifying it as new fuel and lowering radioactivity slightly, its net effect is to greatly increase the volume of radioactive wastes, including that of long-lived wastes.[12]

Table 2: Accumulation of Irradiated Fuel from Commercial Nuclear Plants, 1985 & 1990, with Official Projections for 2000

Country	1985	1990	2000
		(metric tons)	
United States	12,601	21,800	40,400
Canada[1]	9,121	17,700	33,900
Soviet Union	3,700	9,000	30,000
France[2]	2,900	7,300	20,000
Japan	3,600	7,500	18,000
Germany	1,800	3,800	8,950
Sweden	1,330	2,360	5,100
Other[2]	5,939	14,540	36,715
TOTAL[2]	40,991	84,000	193,065

[1] Canadian total is proportionately higher due to its use of natural uranium instead of enriched uranium in its CANDU reactor technology.
[2] France and United Kingdom (listed in "other" category) totals do not include 16,500 tons and 25,000 tons respectively produced by dual-use military and civilian reactors. Nor does the total include these additional 41,500 tons.

Source: Worldwatch Institute, based on sources documented in endnote 10.

Irradiated fuel is only part of the problem. Less intense though still-dangerous "low-level" waste represents a far larger volume. Low-level waste is often described by nuclear authorities as radioactive materials with a half-life of 30 years or less. In fact, debris classified as low-level often contains long-lived materials such as plutonium, technetium, and iodine. Technetium-99 and iodine-129, for example, are routinely found in low-level waste yet have half-lives of 210,000 and 15.8 million years, respectively.[13]

In 1989 alone, more than 76,000 cubic meters of weapons-related low-level waste and 46,000 cubic meters of civilian low-level waste were buried in shallow trenches in the United States. Of the civilian portion, most—about 73 percent of the volume and 95 percent of the radioactivity—came from the nuclear power industry; it included used resins, filter sludge, and discarded equipment. Less than 1 percent of the radioactivity, and 2 percent of the volume, were from medical waste. Because of the wide range of types of low-level wastes, most countries other than the United States classify the longer-lived and more intense varieties as intermediate- or medium-level.[14]

The most voluminous and least concentrated waste comes from uranium mines and mills. Milling removes most uranium from the mined ore, but leaves about 85 percent of the radioactivity in the leftover tailings. A "daughter" of uranium, thorium-230, with a half-life of 77,000 years, predominates in tailings. As it decays, it turns into radium-226, and then radon-222, both potent carcinogens. In the former East Germany, just three of the Wismut uranium mines along the Czechoslovakian border have more than 150 million tons of uranium tailings, with contaminated liquids amounting to millions of tons more. Stabilizing these vast mountains of waste could cost as much as $23 billion. Uranium tailings have also piled up in Australia, Canada, Czechoslovakia, France, Namibia, Niger, South Africa, the Soviet Union, and the United States.[15]

Radioactive waste is also generated by facilities such as uranium conversion and fuel enrichment plants. The Sequoyah Fuels Company plant in Gore, Oklahoma, which converts raw uranium yellowcake to uranium hexafluoride before it is enriched into reactor fuel, has leaked uranium into the soil and groundwater. In October 1991, the U.S. Nuclear Regulatory Commission closed down the plant because of its history of mismanaging hazardous materials.[16]

Nuclear power reactors, like mines and fuel plants, eventually become radioactively contaminated, and—like all other industrial plants—eventually must close. Already, 58 mostly small commercial nuclear reactors have been shut down worldwide, after an average lifetime of just 16 years. The International Atomic Energy Agency (IAEA) estimates that another 60 mostly larger power plants are candidates to be closed by the end of this decade. Dismantling these facilities can produce a greater volume of wastes than operating them: a typical commercial reactor produces 6,200 cubic meters of low-level waste over a 40-year lifetime, and disassembling it creates an additional 15,480 cubic meters of low-level waste. Complete dismantling also requires removal of the irradiated fuel being stored alongside many plants.[17]

The most notorious failure of governments to control nuclear wastes has occurred at U.S. and Soviet military facilities. Weapons manufacturing over the past 50 years at roughly 100 U.S. military sites has led to extreme environmental pollution. According to the U.S. Office of Technology Assessment (OTA), there is "evidence that air, groundwater, surface water, sediments, and soil, as well as vegetation and wildlife, have been contaminated at most, if not all, of the Department of Energy nuclear weapons sites." Contamination from these sites has been found in tumbleweeds, turtles, coyotes, frogs, geese, and shellfish, among other species—and in people.[18]

For decades, in their haste to build weapons, U.S. plant operators vented waste directly into the air or dumped it into the ground, where it found its way into groundwater. Some radioactive wastes ended up in the Columbia River, contaminating shellfish hundreds of kilometers away in the Pacific Ocean. These facilities alone accumulated some 379,000 cubic meters of liquid high-level waste from reprocessing, which was emitting 1.1 billion curies of radioactivity at the end of 1989. The waste is stored in steel tanks at the Hanford Reservation in Washington State and the Savannah River Plant in South Carolina. The tanks have a history of leaking radioactive liquids and accumulating internal buildups of explosive hydrogen gas. Although DOE pledged to clean up these facilities (at a projected cost of $300 billion), its continuing efforts to build a nuclear stockpile have led it to downplay the severity of the problems to government regulators and Congressional overseers.[19]

The situation in the Soviet Union is even worse, according to Thomas Cochran, director of the Natural Resources Defense Council's nuclear project. For Soviet authorities, managing waste once meant dumping it into the nearest body of water. In the case of the Chelyabinsk-40

> **"Lake Karachay became so radioactive that even today a person standing at its shore for an hour would die within weeks."**

weapons facility in the southern Ural Mountains, for example, it was the Techa River that became the chosen repository. In 1951 the Soviet government traced the radioactivity to waters 1,500 kilometers from the plant—in the Arctic Ocean. Weapons builders next pumped waste into Lake Karachay, which Cochran now calls "the most polluted spot on the planet." Lake Karachay became so radioactive that even today a person standing at its shore for an hour would die within weeks due to radiation sickness. In 1967, hot summer winds dried the lake and blew radioactive dust some 75 kilometers away, contaminating 41,000 people.[20]

15

By 1953, the Soviet atomic industry had begun storing reprocessing wastes in steel tanks, although discharges into Lake Karachay did not end until the sixties. The tanks at Chelyabinsk-40 proved to be even less adequate than those at the Hanford and Savannah River plants in the United States. In September 1957, one of the tanks overheated and exploded, spewing a radioactive cloud that contaminated thousands of hectares and required the eventual evacuation of 11,000 people.[21]

This sordid legacy of the Cold War underscores the high costs of mishandling nuclear waste. Estimates for cleaning up the U.S. nuclear weapons plants now run to more than $300 billion. Yet, even such a vast sum may not be enough. According to OTA, "many sites may never be...suitable for unrestricted public access." Uncounted thousands of plant workers and ordinary civilians are likely to die prematurely of cancer and other diseases as a result of their own governments' efforts to build huge nuclear arsenals as quickly as possible. In the military and civilian sectors alike, nuclear wastes have already created permanent sacrifice zones.[22]

Health and Radiation

The problem of radioactive waste from electric power and weapons production goes back fifty years, but human interaction with artificially-induced radioactivity has a record nearly twice as long. In 1904, just nine years after Wilhelm Röntgen's discovery of X-rays, the first technicians started dying from exposure to the mysterious beams. Although some radiologists expressed concern about their colleagues and themselves, most resisted the notion of radiation safety guidelines. Only in 1928 did the Second International Congress of Radiology succeed in setting standards to limit exposure. More than 60 years later, scientists are still struggling to understand just how dangerous radiation really is.[23]

The early X-ray technicians were most concerned about radiation burns and skin ulcers. Since then, however, scientists have found that radiation can lead to cancer, degenerative diseases (such as cataracts), mental retardation, chromosome aberrations, and genetic disorders such as neural tube defects. Radiation also weakens the immune system, allowing other diseases to run their courses unresisted. Damage occurs at the atomic level within individual cells. The energy embodied in the radiation can be transferred to the affected atom, leading to damage, mutation, or destruction of the affected cells. The cumulative effect of cellular change, in turn, is what undermines health. Children and fetuses are particularly susceptible to radiation exposure, because their rapidly dividing cells are more sensitive to damage.[24]

It is difficult to generalize about the health effects of radioactive waste, since each type of radioactive material emits radiation at its own rate. But toxicities can be alarmingly high. In the case of plutonium, for example, it has been noted that less than 150 kilograms, proportionately spread to the lungs of the world's 5.4 billion people, would be enough to cause lung cancer in every one of them. In 1989 alone, U.S. commercial nuclear reactors produced about 67 times that amount—nearly 10 tons—of plutonium, all of which is currently stored in irradiated fuel rods for which no safe disposal method has been found.[25]

The timing and severity of health effects from radiation are closely related to the level of exposure. High doses cause painful death within a few weeks, and intermediate doses have been conclusively shown to cause cancer and other health problems. The health effects of lower doses, particularly those cumulatively received over a period of years, are still debated. However, the prevailing view of scientists appears to be that no dose is innocuous. "There is sufficient evidence that all radiation—however small—presents a risk. There is no threshold" (dose of no consequence), according to Dan Benison of Argentina's Atomic Energy Agency and chairman of the International Commission on Radiological Protection (ICRP), a self-appointed body of radiological and nuclear professionals.[26]

17

The world is continually bathed in "background" radiation, from radon gas seeping from uranium and thorium in the earth's crust and from cosmic rays arriving from outer space. Humans and other life forms evolved over millions of years adapting to these natural levels of radiation, though scientists believe these too have an impact on human health. Radon gas, for example, appears to cause lung cancer. Additional radiation exposure comes from such sources as medical X rays, fallout from the testing of atomic bombs, and the production of nuclear power. For the average person, exposure from these sources is one-fifth that received from background radiation. For individuals working in or living near nuclear installations, it can be much higher.[27]

The most extensive data on the health effects of radiation come from Hiroshima and Nagasaki, Japan, where scientists have been following the health of atomic-bomb survivors since 1950. By comparing the causes of death of those who survived the original blasts and their estimated exposure to radiation, scientists first concluded that lower doses of radiation had minimal health effects. In 1986, however, a reassessment of the original doses found that the victims had been exposed to far less radiation than previously assumed—meaning that the levels of cancer and other health effects attributable to radiation are higher. Using these new data, a 1989 U.S. National Research Council (NRC) committee concluded that an acute dose of radiation is three times likelier to cause cancerous tumors and four times likelier to induce leukemia than was thought 10 years ago.[28]

This information prompted the ICRP, in 1990, again to lower recommended standards for permissible levels of radiation exposure. Indeed, standards have been lowered from 30 centisieverts per year in 1934 to 2 centisieverts in 1990. (A centisievert measures the biological effect on the body of different types of radiation). (See Figure 2.) Some scientists say standards should be decreased even further.[29]

A 1991 study of employees at the U.S. Oak Ridge National Laboratory in Tennessee, which conducts nuclear research and development, suggested that the NRC's estimates of cancer risk may be too low by as much as a factor of 10. It found that leukemia rates were 63 percent higher for nuclear workers exposed to small doses of radiation over long periods than for non-nuclear workers. Likewise, a 1990 report on families of workers at the Sellafield nuclear reprocessing facility in England found

Centisieverts/year
  Whole body

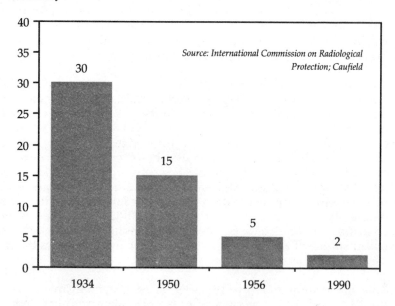

Figure 2. International Recommendations
for Worker Exposure to Radiation

that children are seven to eight times as likely to develop leukemia if their fathers received low-level doses of radiation. The children of Japanese atomic-bomb victims have not experienced similar increases. The reason, hypothesizes the report's author, Dr. Martin Gardner of Britain's University of Southampton, may be that the bomb survivors were subjected to a single, larger dose rather than to the chronic exposure at lower levels of radiation received by nuclear workers. According to John Gofman, a medical doctor and physicist who worked on the Manhattan project during World War II and later founded the Biomedical Research Division at the Lawrence Livermore National Laboratory, the dangers to health from low-level radiation slowly received are 6 to 30 times greater than currently believed by the U.S. National Academy and ICRP.[30]

Other recent investigations have neither confirmed nor disproved a connection between radiation from nuclear facilities and cancer. One study found no radiation-induced health effects on nearby residents six years after the 1979 Three Mile Island accident; but at Oak Ridge it took 26 years for the cancer rate to increase. A U.S. National Cancer Institute report found no association between cancer and nuclear facilities. The data, however, were collected at the county level, which the authors admitted could not spot radiation-induced cancer clusters.[31]

The largest known human experiment with radiation exposure is taking place in Byelorussia, Russia, and the Ukraine, where the largest effects of the estimated 50 million curies released by the 1986 accident at Chernobyl are being felt. Chernobyl's legacy could include hundreds of thousands of additional cancer deaths, yet the Soviet government made no systematic effort to track citizens' cumulative exposure to radiation, or to log health effects. In fact, the effects have been deliberately obscured; a secret Soviet decree prohibited doctors from diagnosing illnesses as radiation-induced. Current estimates of the eventual toll from this disaster range from 14,000 to 475,500 cancer deaths worldwide.[32]

The expanding use of radioactive materials in developing countries is particularly troubling. In 1987, a Brazilian junk worker opened a discarded X-ray machine, extracted a brilliant blue powder—which turned out to be radioactive cesium-137—and distributed it to his family and friends. By the time doctors determined what had happened, four people were fatally contaminated by radiation exposure. Having themselves become a form of radioactive waste, they had to be buried in lead-lined coffins. Another 44 people were hospitalized, suffering from hair loss, vomiting, and other symptoms of radiation sickness. Earlier mishandlings of radioactive wastes had led to deaths in Algeria, Mexico, and Morocco. Third World institutions often are not well prepared to prevent accidental exposures even from the relatively small amounts of radioactive waste generated primarily by research and medical activities.[33]

As both intentional and accidental releases have shown, radiation can be quickly spread through the environment by wind and water. Radioactive waste from Soviet and U.S. military facilities has turned up hundreds of kilometers from its sources, contaminating wildlife, foodstuffs, and people. The fallout from atmospheric atomic bomb testing will eventually cause an estimated 2.4 million cancer deaths worldwide, according to a 1991 study commissioned by the International Physicians for the Prevention of Nuclear War.[34]

Civilian nuclear waste, with its billions of curies of radioactivity, presents a similar threat to the health and safety of future generations. The rupture of a storage facility, failure of a burial site, or a transportation accident could allow radioactive materials to enter water supplies, food products, or even the atmosphere. Indeed, a large, prolonged release of nuclear waste could affect life not just in a local area but over whole regions.

They Call It Disposal

Since the beginning of the nuclear age, there has been no shortage of ideas on how to isolate radioactive waste from the biosphere. Scientists have proposed burying it under Antarctic ice, injecting it into the seabed,

> "The date for a high-level waste burial site was moved to 1989, then to 1998, 2003, and now 2010—a goal that still appears unrealistic."

or hurling it into outer space. But with each proposal has come an array of objections. (See Table 3.) As these have mounted, authorities have increasingly fallen back on the idea of burying radioactive waste hundreds of meters deep in the earth's crust—arguing, as does the U.S. National Research Council, that geologic burial is the "best, safest long-term option."[35]

Most of the countries using nuclear power are now pursuing geologic burial; yet by their own timelines most programs have fallen well behind schedule. In 1975, the United States planned on having a high-level waste burial site operating by 1985. The date was moved to 1989, then to 1998, 2003, and now 2010—a goal that still appears unrealistic. Likewise, Germany expected in the mid-eighties to open its deep burial facility by 1998, but the government waste agency now cites 2008 as the target year. Most other countries currently plan deep geologic burial no sooner than 2020, with a few aiming for even later. (See Table 4.)[36]

The nuclear industry consistently asserts that burying radioactive wastes half a kilometer underground would constitute a technical solution to the problem. According to Jacques de la Ferté, head of external relations and public information at the Nuclear Energy Agency of the Organisation for Economic Co-operation and Development (OECD), the industry has "both the knowledge and the technical resources to dispose of all forms of radioactive waste in satisfactorily safe conditions."[37]

Such blandishments notwithstanding, geologic disposal is nothing more than a calculated risk. Future changes in geology, land use, settlement patterns, and climate all affect the ability to isolate nuclear waste safely. As Stanford University geologist Konrad Krauskopf wrote in *Science* in 1990, "No scientist or engineer can give an absolute guarantee that radioactive waste will not someday leak in dangerous quantities from even the best of repositories."[38]

The concept of geologic burial is fairly straightforward. Engineers would begin by hollowing out a repository roughly half a kilometer below the earth's surface; it would be made up of a broadly dispersed

Table 3: Technical Options for Dealing with Irradiated Fuel

Method	Process	Problems	Status
Antarctica ice burial	Bury waste in ice cap	Prohibited by international law; low recovery potential, and concern over catastrophic failure	Abandoned
Geologic burial	Bury waste in mined repository	Uncertainty of long-term geology, groundwater flows, and human intrusion	Under active study by most nuclear countries as favored strategy
Seabed burial	Bury waste in deep ocean sediments	May violate international law; transport concerns; nonretrievable	Under active study by consortium of 10 countries
Space disposal	Send waste into solar orbit beyond earth's gravity	Potential launch failure could contaminate whole planet; very expensive	Abandoned
Long-term storage	Store waste indefinitely in specially constructed buildings	Dependent on human institutions to monitor and control access to waste for long time period	Not actively being studied by governments, though proposed by non-govt'l groups
Reprocessing	Chemically separate uranium and plutonium from irradiated fuel	Increases volume of waste by 160 fold; high cost; increases risk of nuclear weapons proliferation	Commercially underway in four countries; total of 16 countries have reprocessed or plan to reprocess irradiated fuel
Transmutation	Convert waste to shorter-lived isotopes through neutron bombardment	Technically uncertain whether waste stream would be reduced; very expensive	Under active study by United States, Japan, Soviet Union, and France.

Source: Worldwatch Institute, based on sources documented in endnote 35.

series of rooms from which thermally hot waste would be placed in holes drilled in granite, clay, volcanic tuff, or salt. Waste would be transported to the burial site in trucks, trains, or ships. Technicians would package it in specially constructed containers made of stainless steel or other metal. Once placed in the rock, the containers would be surrounded by an impermeable material such as clay to retard groundwater movement, then sealed with cement. When the repository is full, it would also be sealed off from the surface. Finally, workers would erect some everlasting sign-post to the future—in one DOE proposal, a colossal nuclear stonehenge—warning generations millennia hence of the deadly radioactivity entombed below. The rationale for such warning is that the durability of the tomb is only as good as the durability of the information needed to protect it from disturbance—and human institutions have no proven capability to protect information over hundreds of generations.[39]

The cost of building such mausoleums is as uncertain as their security. Like the projected dates of completion, and in keeping with the tradition of nuclear construction projects, estimates of needed expenditures keep rising. In the United States, the expected cost of burying irradiated fuel has risen 80 percent just since 1983, with the bill for a single site holding 96,000 tons of irradiated fuel and high-level waste now projected at $36 billion.[40]

Knowledge about deep geology comes principally from mining, an activity aimed at extracting valuable mineral resources in a short period. With deep burial of nuclear waste, the task is quite different: to provide adequate isolation for thousands of years. "The technical problem is not of digging a hole in the ground; it's of forecasting the unknown," says Scott Saleska, staff scientist at the Institute for Energy and Environmental Research in Maryland.[41]

According to a 1990 NRC report on radioactive waste disposal, the needed long-term quantitative predictions stretch the limits of human understanding in several areas of geology and groundwater movement and chemistry. "Studies done over the past two decades have led to the realization that the phenomena are more complicated than had been thought," notes the report. "Rather than decreasing our uncertainty, this

Table 4: Selected Country Programs For High-Level Waste Burial

Country	Earliest Planned Year	Status of Program
Argentina	2040	Granite site at Gastre, Chubut, selected.
Belgium	2020	Underground laboratory in clay at Mol.
Canada	2020	Independent commission conducting four-year study of government plan to bury irradiated fuel in granite at yet-to-be-identified site.
China	none announced	Irradiated fuel to be reprocessed; Gobi desert sites under investigation.
Finland	2020	Field studies being conducted; final site selection due in 2000.
France	2010	Three sites to be selected and studied; final site not to be selected until 2006.
Germany	2008	Gorleben salt dome sole site to be studied.
India	2010	Irradiated fuel to be reprocessed, waste stored for twenty years, then buried in yet-to-be-identified granite site.
Italy	2040	Irradiated fuel to be reprocessed, and waste stored for 50-60 years before burial in clay or granite.

Country	Earliest Planned Year	Status of Program
Japan	2020	Limited site studies. Cooperative program with China to build underground research facility.
Netherlands	2040	Interim storage of reprocessing waste for 50-100 years before eventual burial, possibly sub-seabed or in another country.
Soviet Union	none announced	Eight sites being studied for deep geologic disposal.
Spain	2020	Burial in unidentified clay, granite, or salt formation.
Sweden	2020	Granite site to be selected in 1997; evaluation studies under way at Äspö site near Oskarshamn nuclear complex.
Switzerland	2020	Burial in granite or sedimentary formation at yet-to-be-identified site.
United States	2010	Yucca Mountain, Nevada, site to be studied, and if approved, receive 70,000 tons of waste.
United Kingdom	2030	Fifty-year storage approved in 1982; explore options including sub-seabed burial.

Source: Worldwatch Institute, based on sources documented in endnote 36.

line of research has increased the number of ways in which we know that we are uncertain."[42]

In the United States and Germany, where specific burial sites have been selected for assessment and preparation, the work to date has raised more questions than answers about the nature of geologic repositories. German planners have targeted the Gorleben salt dome, 140 kilometers from Hannover in northern Germany, to house the country's high-level waste from reprocessed irradiated fuel by 2008. The DOE is focusing on two locations: Nevada's Yucca Mountain for high-level waste, and the Waste Isolation Pilot Plant (WIPP) in southeastern New Mexico for long-lived transuranic waste from the U.S. nuclear weapons program. Construction of some burial rooms has already been completed at WIPP, which could become the world's first deep repository for nuclear waste.[43]

As the experience with these sites has shown, the most intractable problem with deep burial is water. In Germany, groundwater from neighboring sand and gravel layers is eroding the salt that makes up the Gorleben dome. In New Mexico, WIPP's burial rooms—more than 600 meters deep in a mass of salt—are sandwiched between two bodies of water. Below the repository is a reservoir of brine (salt water), and above is a circulating groundwater system that feeds the Pecos River. The salt rooms at WIPP were expected to be dry, but brine is constantly seeping through the walls. Corrosive groundwater could easily eat away steel waste drums and create a radioactive slurry.[44]

This danger is compounded by the fact that 60 percent of the transuranic waste at WIPP also contains hazardous chemicals such as flammable solvents. This "mixed waste" gives off gases, including explosive hydrogen, that could send a plume of radioactive slurry into the aquifer above it. The aquifer, in turn, could contaminate—via the Pecos—much of the irrigation water on which the huge Rio Grande Delta agricultural economy depends.[45]

Groundwater conditions at the U.S. site at Yucca Mountain, a barren, flat-topped ridge about 160 kilometers north of Las Vegas, are also rais-

"The salt rooms at WIPP were expected to be dry, but brine is constantly seeping through the walls."

ing concerns. According to the current plan, the waste deposited in Yucca Mountain's volcanic tuff would stay dry because the storerooms would be located more than 300 meters above the present water table, and because percolation from the surface under current climatic conditions is minimal. But critics, led by DOE geologist Jerry Szymanski, believe that an earthquake at Yucca Mountain, which is crisscrossed with more than 30 seismic faults, could dramatically raise the water table. Others disagree. But if water came in contact with hot radioactive wastes, the resulting steam explosions could burst open the containers and rapidly spread their radioactive contents. "You flood that thing and you could blow the top off the mountain. At the very least, the radioactive material would go into the groundwater and spread to Death Valley, where there are hot springs all over the place," says University of Colorado geophysicist Charles Archambeau.[46]

27

Human actions, as well as geological ones, could disrupt the planned repositories. Prospectors and miners could be attracted by gold and silver deposits near Yucca Mountain, or by natural gas and potash deposits in the area around New Mexico's WIPP. Salt domes like Gorleben are also targets for natural gas exploration.[47]

Scientific knowledge simply may not yet be up to the task of isolating nuclear waste for eons. In 1990, scientists discovered that a volcano 20 kilometers from Yucca Mountain erupted within the last 20,000 years—not 270,000 years ago, as they had earlier surmised. Volcanic activity could easily resume in the area before Yucca Mountain's intended lethal stockpile is inert. It is worth remembering that less than 10,000 years ago, volcanoes were erupting in what is now central France; that the English Channel did not exist 7,000 years ago; and that much of the Sahara was fertile just 5,000 years ago. Only a clairvoyant could choose an inviolable, permanent hiding place for nuclear waste.[48]

Technical Fixes?

In 1972, nuclear pioneer and then Oak Ridge National Laboratory director Alvin Weinberg, sensing the inadequacy of geologic burial, suggested a startling alternative: indefinite storage in surface facilities that would be guarded and tended by what Weinberg called a "nuclear priesthood." Such an endeavor, however, would have to overcome the fragility of human institutions, as much of the waste is likely to remain dangerous for a period far longer than that of recorded human history. Furthermore, above-ground storage increases the potential for breaching of a storage facility, whether accidental or intentional.[49]

In 1958, a German patent was issued for a proposal to place high-level waste in the icesheets of Antarctica. However, scientific and political questions soon arose. Icesheets periodically surge forward, and hot waste could melt the ice, facilitating such movements. Radioactive wastes could then end up in the ocean. Difficulties would also arise over transporting waste across ice-filled and stormy seas. The scientific uncertainties, along with the anticipated difficulty of achieving international approval to abandon waste on international territory, effectively killed the idea. (Indeed, a treaty signed in 1991 puts Antarctica off limits for conventional mining for the next 50 years.)[50]

An idea given more serious consideration was that of rocketing radioactive waste into outer space. One concept, formulated by the U.S. National Aeronautics and Space Administration (NASA), would have employed the space shuttle to launch high-level waste into a solar orbit between those of the earth and Venus. The idea, however, suffered a decisive setback when the Space Shuttle Challenger exploded during take-off in 1986. In 1991, the idea quietly resurfaced. But even if technically feasible, outer-space disposal appears to be prohibitively expensive—and out of the question for the bulk of nuclear waste.[51]

Burial in the deep ocean seabed is still actively being investigated by international agencies and governments. Waste would be sunk 50 meters into the bottom sediments in areas of the deep ocean floor that have been stable for up to half a million years. When the substances leak out, as they eventually would, scientists believe that the radionuclides would bind with clay sediments, locking them in place. Poor circulation of deep bottom waters of the oceans would help isolate the radioactivity.[52]

One drawback to ocean burial is that it nearly eliminates the possibility of retrieval, should anything go wrong. Political problems would also be likely, particularly since earlier dumping by the United States and several European countries at over fifty sites in the northern Pacific and Atlantic Oceans is seen as posing a threat to local ocean ecosystems. Only now are studies being conducted to determine the effects of haphazard dumping by the United States between 1946 and 1970. The first site to be investigated is near the Farallon Islands, 30 miles west of San Francisco and home to a major fishery and the largest population of breeding seabirds south of Alaska.[53]

Existing treaties could block further sea-burial efforts. The London Dumping Convention, signed in 1972, prohibits high-level waste dumping in oceans, and the convention's parties agreed to a moratorium on dumping low-level wastes into the seas in 1983. In 1990, by a vote of 29 to 4, the convention parties approved a non-binding resolution to prohibit the scuttling of retired nuclear submarines at sea. While it is unclear whether the convention would cover attempts to carefully bury waste *below* the ocean floor, they would certainly meet strong opposition from many nations. Use of any ocean bed as a nuclear septic field would be unlikely to succeed without a prodigious effort to win international backing.[54]

Another strategy to which authorities have been attracted is that of reprocessing irradiated fuel. First developed in the United States and other countries under the aegis of secret military programs to extract plutonium and unfissioned uranium from irradiated uranium fuel, reprocessing was then promoted in civilian nuclear industries as a way to overcome predicted shortages of natural uranium. Part of the rationale for this strategy is that reprocessing, along with fast breeder reactors that operate on plutonium, could provide enough fuel for nuclear fission to last for centuries. As it turns out, however, the need for fuel is no longer so urgent. Due to slower growth than expected in the nuclear power industry, along with expanded discoveries of uranium, these resources will not be a limiting factor before 2030, according to the IAEA and the NEA.[55]

Despite the current glut of uranium, France, India, Japan, the Soviet Union, and the United Kingdom remain committed to reprocessing commercial irradiated fuel. The United States, on the other hand, saw its only commercial reprocessing plant (at West Valley, New York) close in 1972 due to economic and technical problems. Waste still located at the facility will cost an estimated $3.4 billion to contain. In 1977, President Jimmy Carter indefinitely deferred U.S. efforts to reprocess commercial irradiated fuel in order to slow down the international proliferation of nuclear weapons materials. While his order was rescinded by the Reagan administration, neither the U.S. government nor the nuclear industry have made a real effort to revive commercial reprocessing. Despite Carter's efforts to slow the spread of reprocessing, many countries that do not have reprocessing plants send their irradiated fuel to Britain and France. This has led to extensive international movement of irradiated fuel—and will soon lead to an escalation in the always-risky transport of radioactive waste and plutonium. Sixteen countries have or plan to have their fuel reprocessed, according to the IAEA.[56]

Because it reduces the most potent component of the waste only by increasing the production of intermediate and low-level varieties (including those that are long-lived), reprocessing irradiated fuel greatly increases the total volume of waste that needs to be managed. Even by industry estimates, reprocessing multiplies the quantity of wastes requiring long-term isolation from the biosphere nearly ten-fold. It increases the quantity of low-level wastes by an even greater amount. Overall, reprocessing expands the volume of radioactive waste 160-fold over the original irradiated fuel, according to the British Central Electricity Generating Board.[57]

Nor does economics favor reprocessing. The new British facility being built at Sellafield quadrupled in cost during the past thirteen years, and a recent analysis by Colin Sweet, an independent energy analyst in Wales, found that uranium derived from reprocessing irradiated fuel will be 18 times more expensive than from mined uranium. And economics doesn't justify the use of reprocessing for waste disposal any more than it justifies reprocessing for producing fuel. A German study

"Even by industry estimates, reprocessing
multiplies the quantity of wastes requiring
long-term isolation nearly ten-fold."

found that burying irradiated fuel would be 30 to 40 percent less expensive than reprocessing it and then burying the waste. Rising costs, along with public protests over safety, led the Germans in 1989 to cancel their plans to build a reprocessing facility at Wackersdorf.[58]

Countries utilizing reprocessing are uncertain what to do with all the plutonium being separated, since efforts to rely on plutonium-fueled fast breeder reactors have been shelved due to high costs and technical failures. The only large breeder, the 1,200 megawatt French Superphénix, reportedly cost three times as much to build as a standard light-water reactor and has had such an abysmal operating record—on-line less than 40 percent of the time—that the French government is debating whether to shut it down permanently. India's trouble-plagued prototype breeder has reportedly not operated for more than a few minutes at a time, while Soviet efforts to build two large breeder reactors have been stymied by economic problems and local opposition.[59]

A secret French government report, leaked to the press in March 1990, admitted that reprocessing was producing far more plutonium than could be used by the French nuclear industry. Along with Germany and Japan, France plans to expand experimental programs to burn up separated plutonium mixed with uranium (known as mixed-oxide, or MOX, fuel) in existing light-water reactors. Yet the IAEA expects that less than 20 percent of the reprocessed plutonium will be used in MOX fuel. What happens to the remaining 30-odd tons that could be produced annually by 1995—enough plutonium for nearly 4,000 atomic bombs—is unknown. The potential for diverting plutonium through hijacking or theft obviously increases, as more of it moves into circulation between countries as distant as England and Japan.[60]

While reprocessing is not necessary for obtaining nuclear weapons-grade material, it can make such acquisition much easier. The plutonium for India's 1974 nuclear explosion, for example, was separated at the country's first reprocessing facility, allowing India to join China, France, the Soviet Union, the United Kingdom, and the United States as an avowed nuclear power. A number of other countries, including Israel,

Table 5: **Countries with Nuclear Weapons Programs.**

Recognized Nuclear Weapons Countries	Countries Believed To Possess Nuclear Weapons	Countries Believed to have Nuclear Weapons Programs[1]
China	Israel	Algeria
France	India	Argentina
Soviet Union	Pakistan	Brazil
United Kingdom	South Africa	Iran
United States		Iraq
		Libya
		North Korea
		South Korea
		Sweden
		Taiwan

[1]Current or historical.

Sources: George W. Rathjens and Marvin M. Miller, "Nuclear Proliferation After the Cold War," *Technology Review*, August/September 1991; Leonard Spector, with Jacqueline R. Smith, *Nuclear Ambitions: The Spread of Nuclear Weapons 1989-1990* (Boulder, Colo: Westview Press, 1990); R. Jeffrey Smith, "Officials Say Iran Is Seeking Nuclear Weapons Capability," *Washington Post*, October 30, 1991; Elaine Sciolino and Eric Schmitt, "Algerian Reactor: A Chinese Export," *New York Times*; November 15, 1991.

Pakistan, and South Africa, are believed capable of quickly assembling atomic weapons today.[61] (See Table 5.)

Another ten countries have shown interest or partial success in producing nuclear weapons. The apparent ease with which Iraq moved toward manufacturing nuclear weapons—including separating small amounts of plutonium from irradiated fuel—while in apparent compliance with the Non-Proliferation Treaty and under IAEA safeguards, illustrates the shortfall of current efforts to prevent proliferation. Waste strategies that make it easier for countries to obtain nuclear weapons material will only accelerate proliferation.[62]

David Albright, a physicist with Friends of the Earth, states that "current plans to proceed with reprocessing seem to be driven mostly by inertia, old habits of mind, and political expediency, including the desire of governments and utilities to postpone decisions on radioactive waste disposal." Reprocessing, in other words, has become a form of interim storage for nations not knowing what to do with irradiated fuel. Countries such as Germany and Switzerland, for example, pay to send their irradiated fuel to the United Kingdom or France, and so delay the return of high-level waste for some twenty years.[63]

Across Europe, reprocessing is getting increased scrutiny. In February, 1991, a study by the French College for the Prevention of Technical Risks, an independent government office, called for a reconsideration of reprocessing and a reexamination of direct burial of irradiated fuel. Scottish Nuclear, the state-owned company that runs four nuclear plants in Scotland, has decided for economic reasons to forego new contracts for reprocessing its irradiated fuel, and plans on long-term storage instead. The Germans are reconsidering their national law that requires reprocessing of irradiated fuel. And in June 1991, the European Parliament approved two non-binding resolutions calling for the end of reprocessing. However, England, France, Japan, and India are still planning to proceed with their programs, which involve opening major new reprocessing plants in the next few years—and more shipments of irradiated fuel, plutonium, and radioactive waste around the globe.[64]

Despite its drawbacks, reprocessing is still seen by some nuclear proponents as integral to the "true solution" to the waste problem. The argument is that plutonium and uranium from reprocessed irradiated fuel could be burned up in light-water reactors or in fast breeder reactors in a continuous cycle. Ultimately, residual long-lived waste from reprocessing would be transformed into relatively short-lived isotopes, through a process almost wistfully known as "transmutation."[65]

Over the years, various concepts of transmutation have been advanced, yet none have survived serious examination so far. A novel method proposed by researchers at the Los Alamos National Laboratory in New

Mexico uses a particle accelerator to shoot sub-atomic particles—neutrons—into the radioactive atoms. When the waste atom absorbs a neutron, it is converted to a different atom, one that could be less long-lived or even non-radioactive. Key to the Los Alamos method is an improved technique to separate the radioactive elements into waste streams of selected isotopes.[66]

But transmutation could be as fanciful as turning lead into gold—theoretically possible, but economically and technically unreachable. The OECD's Nuclear Energy Agency (NEA) believes that transmutation "will not be feasible in the near future." Even if feasible, transmutation could end up increasing the amount of waste due to handling and processing, according to Thomas Pigford, a nuclear engineering professor at the University of California, Berkeley. It might also be difficult to prevent the extra neutrons from turning non-radioactive substances into radioactive ones, as happens in the core of nuclear reactors. Furthermore, the amount of energy needed to transmute nuclear waste may exceed that produced by the irradiated fuel. And even proponents say that transmutation would not eliminate the need for disposal of the remaining wastes.[67]

Despite the many problems with the technological fixes proposed to date, there have been successes. Safer methods of temporarily storing radioactive waste, particularly irradiated fuel, reduce the chance that radioactivity will spread to the environment. The U.S. Nuclear Regulatory Commission says dry casks can be safe for interim storage for at least 100 years. Such storage technologies could also reduce the transport of irradiated fuel for reprocessing, eliminating another weak link in the nuclear fuel cycle. Future technological advances will be needed not only to contain the growing waste stockpile, but also to clean up the contamination that has already entered the biosphere.[68]

The Politics of Nuclear Waste

Both government and industry have long ignored warnings about radioactive wastes. In 1951, Harvard University president and former

"The ground around the site was a "Swiss cheese"
of old oil and gas wells through which
groundwater might seep."

Manhattan Project administrator James B. Conant spoke publicly of wastes that would last for generations. In 1957, a U.S. National Academy of Sciences (NAS) panel cautioned that "unlike the disposal of any other type of waste, the hazard related to radioactive wastes is so great that no element of doubt should be allowed to exist regarding safety." In 1960, another Academy committee urged that the waste issue be resolved *before* licensing new nuclear facilities.[69]

35

Yet such recommendations fell on deaf ears, and one country after another plunged ahead with building nuclear power plants. Government bureaucrats and industry spokespeople assured the public that nuclear waste could be dealt with. However, early failures of waste storage and burial practices resulted in growing mistrust of the secretive government nuclear agencies that were responsible. Two of the six shallow burial sites for commercial low-level radioactive waste in the United States, for example, have leaked, and three have been closed. Trust has also faded as the public has come to view government agencies as more interested in encouraging the growth of nuclear power than in resolving the waste problem. The result has been grassroots opposition to nearly any attempt to develop radioactive waste facilities.[70]

The United States has perhaps the most dismal history of mismanaging waste issues. Over 100 reactors now have operating licenses, accounting for about 20 percent of the country's electricity; yet from the fifties onward, nuclear waste problems have been swept under the rug by the U.S. Atomic Energy Commission and its successors. Only following a stinging 1966 NAS critique of the AEC's waste policy (suppressed by the AEC until Congress demanded its release in 1970), and a 1969 accident at the U.S. government's bomb-making facility at Rocky Flats, Colorado, did the AEC concoct a hasty attempt to solve the problem by planning to bury nuclear waste in a salt formation in Lyons, Kansas. By 1973 the AEC was forced to cancel the site because in its haste it had overlooked serious technical problems—such as the fact that the ground around the site was a "Swiss cheese" of old oil and gas wells through which groundwater might seep. The Lyons failure set off a decade of wandering from site to site, and of growing opposition from increasingly apprehensive states.[71]

In 1976, the California legislature approved a moratorium on building new nuclear power plants until the federal government approved a "demonstrated technology or means for the disposal of high-level nuclear waste." Seven states soon passed similar legislation that tied future nuclear power development to solution of the waste problem. The growth of nuclear power seemed threatened, and the nuclear industry pushed the AEC's successor, the Department of Energy, to bury waste quickly.[72]

DOE had no better success in finding a state amenable to housing the nation's waste. The department's repeated failures prompted the Congress to pass the Nuclear Waste Policy Act of 1982. A product of byzantine political bargaining, the law required DOE to develop two high-level repositories, one in the western part of the country and the other in the east. The department was hampered, however, by an unrealistic timetable and its own insistence on considering sites that were clearly unacceptable. In one case, DOE ignored findings by government scientists that a proposed repository in Hanford, Washington could leak in short order. As DOE failed to gain public confidence, the whole process became embroiled in political conflicts at the state level. Finally, when eastern states forced the cancellation of an eastern repository in 1986, the legislation fell apart. With the whole program in jeopardy, and over the strong objections of the Nevada delegation, Congress ordered DOE to study just one site—adjacent to the federal government's nuclear test area near Yucca Mountain.[73]

The federal government appears determined to saddle Nevada with the country's waste. The state has vigorously sought the disqualification of the site, and it challenges DOE's ability to conduct research objectively, given that Yucca Mountain is the only site being investigated. So vehement are the objections of Nevadans that the state legislature in 1989 approved a bill prohibiting anyone from storing high-level waste in the state. Although the law has not been tested yet, it appears to violate the federal preemption on nuclear matters. Former Nuclear Regulatory Commissioner Victor Gilinsky describes Yucca Mountain as a "political dead-end."[74]

Transportation is another concern surrounding the Nevada site. An average load of irradiated fuel from an eastern reactor would travel 2,000 miles. States along the potential transportation route, including Colorado and Nebraska, have already expressed their citizens' anxiety over radioactive releases either from accidents or from normal emissions from waste containers. DOE, however, will consider transportation problems only after the site is judged to be technically sound.[75]

Delays have led to postponing waste burial in Yucca Mountain until 2010. The U.S. General Accounting Office faults DOE's own ineptness for the delays, but the department blames Nevada and is strongly supporting efforts in Congress to reduce the state's participation in the review process. DOE's program has also been criticized by Federal judge Reginald Gibson, who remarked that there was a strong suggestion of possible "impropriety and influence" in DOE's awarding of a $1-billion repository design contract for Yucca Mountain. In addition, Yucca Mountain is in a region claimed by the Western Shoshone Indians. Faced with a long list of technical difficulties and firm political opposition, Yucca Mountain has a long way to go before becoming the country's burial site for high-level waste.[76]

France is second only to the United States in production of nuclear electricity, as the result of an ambitious construction program over the past 15 years. Since 1987, however, only one order has been placed for a new reactor, and the program has faced serious challenges over safety, economics, and waste management. In 1987, the French waste agency, ANDRA, announced four potential sites for burying high-level radioactive wastes. Local government officials, disturbed that they had not been consulted or forewarned, joined with farmers and environmentalists to organize an effective campaign to stop the research program. Blockades obstructed government technicians, and geologic survey work proceeded only with police protection. In January 1990, in one of the country's largest anti-nuclear demonstrations since the late seventies, 15,000 people marched in Angers against the Maine-et-Loire site in west central France. By February, Prime Minister Rocard had imposed a nationwide moratorium on further work at the four sites, providing the government a cooling-off period to reevaluate its waste program and try again to win public support.[77]

In the same year, however, the French government's credibility suffered a further setback when an independent organization, the Commission for Independent Research and Information on Radioactivity (CRII-RAD), discovered plutonium and other radionuclides at the Saint-Aubin sewage dump near Paris. While the government acknowledged that the facility had earlier been used to store radioactive waste, it initially denied that plutonium had ever been there, and claimed that Saint-Aubin had been decontaminated of other radioactive wastes by 1979. It later admitted that plutonium was not only present as the critics had claimed, but was there in levels over twice as high as CRII-RAD had first calculated.[78]

In an attempt to get the waste program back on track, the French parliament approved a new plan in June 1991. ANDRA, which Parliament made autonomous from the country's Atomic Energy Commission, will reduce the number of sites to be investigated from four to two. Still, no sites have been named. The government says the selection process will be more open than the previous one, with independent experts consulted. Officials hope to convince local communities to accept site investigations by paying local treasuries up to $9 million a year for "the psychological inconvenience" of being studied, according to then Industry Minister Roger Fauroux. Any decision on a final burial site will be delayed for 15 years, and then Parliament will make the final decision. In the meantime, the country's high-level waste inventory will more than triple.[79]

In Germany, the controversy over radioactive waste mirrors that surrounding nuclear reactor construction, which has come to a standstill with 21 plants built. Local opposition to any nuclear project, and an inability of the major political parties to agree on nuclear policy, appear deeply entrenched. In 1991, strong public opposition led to the permanent closure of a new nuclear reactor, the Kalkar breeder plant. It also stymied plans to build new plants in the former East Germany, where all existing reactors were closed in 1990. The waste program has become caught up in a similar debate, with the Lower Saxony government joining local people who vigorously oppose federal plans to build nuclear waste burial sites at Konrad and Gorleben.[80]

Since the sixties, German policy has been to bury deeply all wastes, but the early failure to gain public confidence has led to conflicts. The low-level site at Asse was closed in 1978, due to local opposition and stability problems. Public opposition and technical uncertainties also delayed work at the replacement site, the abandoned iron ore mine at Schacht Konrad. As a result, low-level wastes have been piling up, with some 50,000 cubic meters currently stored at temporary facilities and reactor sites. Recently, work on Konrad continued only following judicial rulings against Lower Saxony. Hopes that the low-level waste dump at Morsleben in the former East Germany could handle the country's waste were dampened when the facility was closed by German officials in early 1991.[81]

Public opposition has also frustrated Germany's attempt to develop a high-level waste burial site. In 1976, the federal government's first three investigation sites in Lower Saxony created such an uproar among local farmers and students that the state government rejected them. The following year, the federal government selected a salt dome at Gorleben, in Lower Saxony, along what was then the East German border. Large protests erupted even before the official announcement; 2,500 people took over the drilling site for three months before police hauled them off and set up a secure camp from which scientific work could be conducted. Although the federal government has put all its bets on Gorleben, continuing technical problems and protests make plans to bury waste by 2008 highly unrealistic. Critics have warned that the site's geology is unstable, and one worker was killed by collapsing rock during a 1987 drilling accident, further eroding public confidence; but the government waste agency continues its work at the site.[82]

In Sweden, nuclear issues have been erupting since the seventies, when two governments were thrown out of office over nuclear energy policies.

In 1977, Parliament passed the Stipulation Act, requiring utilities to demonstrate the technical feasibility of safe high-level waste disposal. Only following a national referendum in 1980, and Parliament's subsequent decision to limit the number of reactors in the country to 12 and to phase them out by 2010, was the country able to focus on the waste issue. One immediate dividend from the agreed phase-out was a clarification of exactly how much waste would eventually need to be dealt with: 7,750 tons of irradiated fuel, 90,000 cubic meters of low-level and intermediate-level waste, and 115,000 cubic meters of "decommissioning" waste from the dismantling of nuclear plants.[83]

Sweden's high-level waste program has won international praise for relying not simply on deep burial but on a system of redundant engineering barriers, starting with a corrosion-resistant copper waste canister with four-inch thick walls that has an estimated lifetime of 100,000 years or longer if undisturbed. The country has also avoided major siting problems by locating both of its operating waste facilities next to power plants. Irradiated fuel is being temporarily stored 30 meters underground at the CLAB facility alongside the Oskarshamn nuclear power plant. To reduce transportation controversies, irradiated fuel is shipped to CLAB from reactors by a special boat rather than by truck or rail. Since 1988, low-level wastes have been placed in a mined cavern 50 meters below the Baltic Sea near the Forsmark nuclear power plant.[84]

Swedish public support has not come as easily for deep burial of irradiated fuel as for low-level waste, however. Protests halted attempts to site a permanent high-level burial facility 10 years ago. Efforts to explore other sites have met determined local opposition since then, even though opinion polls indicate that most Swedes would in principle accept a disposal site in their community. The government is likely to focus once again on a reactor site for its final irradiated fuel repository.[85]

Japan has one of the world's few remaining nuclear construction programs, with work progressing on some 11 new reactors to add to the existing 41. But public opposition to nuclear facilities has grown over the past decade, particularly since the Chernobyl accident and a series of

> "Opposition to the Rokkasho high-level waste storage site was quieted, however, by the distribution of large amounts of money."

recent mishaps. As a consequence of the Japanese tradition of not siting industrial facilities without local government support, no new reactor sites have been approved since 1986, and attempts to locate waste facilities have proved equally controversial.[86]

Serious efforts to deal with nuclear waste in Japan did not begin in earnest until the early eighties, when the country started switching on a large number of reactors ordered in the seventies, and waste began piling up. The nuclear industry selected the village of Rokkasho on the northern tip of Honshu Island to house a complex for reprocessing, high-level waste storage, and low-level waste burial. Local opposition to the project has been strong, with candidates opposing the facility winning mayoral and legislative elections in 1989 and 1990, and jeopardizing the entire project.[87]

Opposition to Rokkasho was quieted, however, by the distribution of large amounts of money. In return for local acceptance of the project, the Japanese government offered $120 million in subsidies to the village—equivalent to about $10,000 for each resident. An additional $120 million was offered to surrounding villages. The amounts are small when compared with the project's price tag of $9 billion, but Rokkasho is located in Japan's second poorest prefecture, and the offer appears to have had its intended effect. In February 1991, the incumbent governor, a supporter of the nuclear project, won reelection over two anti-nuclear candidates with 44 percent of the vote.[88]

While Rokkasho is expected to be the temporary storage place for high-level waste being returned from reprocessors in Europe as early as 1993, attempts to locate a final burial site have so far been thwarted by public opposition. In 1984, planners again looked to a poor region of the country, selecting an amenable village, Horonobe, near the northern tip of Hokkaido Island. But opposition from the Hokkaido Prefecture governor and diet and from nearby villages and farmers has blocked government plans to construct a waste storage and underground research facility in Horonobe. Meanwhile, attempts to undertake exploratory drilling in other parts of the country have also faced protests.[89]

There are signs that Japan is now looking beyond its borders for a high-level waste disposal site. Since 1984, China has shown interest in importing irradiated fuel or waste for a fee, or in return for assistance with its own fledgling nuclear program. In November 1990, China and Japan agreed to build an underground facility in China's Shanxi province, where research is to be undertaken on high-level waste burial.[90]

In the Soviet Union, anti-nuclear groups have proliferated since the Chernobyl accident, bringing to a halt not only the country's nuclear construction program but also attempts to deal with nuclear wastes. Trust between the public and nuclear authorities is at a low point, as the Soviet press has been inundated with reports of "radiation disaster zones" unrelated to Chernobyl.[91]

The Soviet Ministry of Atomic Power and Industry reports that it is looking for burial sites near the Chelyabinsk site in the southern Urals, home to the country's main reprocessing facility, but is encountering opposition from local people. Efforts to build a major reprocessing plant at Krasnoyarsk in Siberia have been postponed, partly as a result of a petition delivered to local authorities in June 1989, signed by 60,000 people protesting reprocessing and waste disposal plans. Further complicating the picture is a 1990 law passed by the Russian parliament that prohibits the burial of radioactive wastes from other Soviet republics or foreign countries. The Soviets also face criticism from Norway and Finland concerning waste storage facilities in the western part of the country.[92]

Bulgaria, Czechoslovakia, and Hungary have in the past returned irradiated fuel to the Soviet Union for reprocessing, without having to take back the waste. Since the late eighties, however, the Soviets have insisted on charging for a service they previously provided for free, and shipments from Central Europe have ceased. Irradiated fuel is now building up in temporary storage facilities that will be full in two to five years. The three countries are planning to expand storage capabilities, but eventually they will need to decide what is to become of the waste. Furthermore, public confidence has ebbed, with past burials of radioactive materials already coming back to haunt the new democracies.[93]

Radioactive waste has created serious political problems elsewhere, as well. In South Korea, the 1988 discovery of illegally dumped radioactive wastes in Changan Village near the Kori nuclear power plant spawned the country's anti-nuclear movement. Since then, protests—some of them violent—have stopped government workers from developing low-level waste sites on two islands. The government now hopes to bury radioactive substances on uninhabited islands by 1995. No firm plans beyond temporary above-ground storage have been revealed for high-level wastes.[94]

In Argentina, too, public pressure has forced the government to renounce plans to quickly build a deep repository for high-level waste from the country's two operating reactors. And in Taiwan, government efforts to handle irradiated fuel from the country's six nuclear plants have advanced marginally: a research program, including geologic studies, has been initiated in the past five years, but no sites have been chosen yet. Meanwhile, aboriginal Yami people have protested the government's nuclear dump for low-level waste on Orchid Island. And Taiwanese concern over radioactive waste is contributing to the growing public opposition to building more reactors.[95]

Among the developing countries, India has the most ambitious construction program, with 13 relatively small reactors planned or under construction to add to the seven already built. A budding anti-nuclear movement confronts an entrenched bureaucracy accustomed to secrecy, but it has already succeeded in frustrating the country's nuclear planners. Citizen opposition stalled plans to build two reactors near Hyderabad. Efforts to site a nuclear waste repository, apparently not yet commenced, could be hindered by the country's high population density. Meanwhile, the government devotes less than 1 percent of its substantial nuclear power budget to waste management.[96]

Worldwide, it's questionable whether any government has the political capacity to build and operate nuclear waste repositories, so strong is the public opposition that has entangled nuclear power. So far, governments have made short-term decisions on waste while leaving their long-term plans vague, hoping to muddle through. While this makeshift strategy may calm immediate concerns, it has not led societies closer to a solution to the waste dilemma; nor has it bridged the gap between governments and their citizens. If governments continue in their attempts to site waste facilities over public opposition, that political gulf may grow wider still.

Beyond Illusion

François Chenevier, director of the French nuclear waste agency ANDRA, recently admonished that "it would be irresponsible for us to benefit from nuclear power and leave it to later generations to deal with the waste." Yet that has already occurred. While nuclear reactors generate electricity for 25-40 years, their radioactive legacy will remain for thousands of centuries. So far, none of the waste has been dealt with in a manner that will withstand this test of time.[97]

The problem of radioactive waste, however, can never be "solved" in the normal fashion. Waste cannot be destroyed, nor can scientists prove that it will stay out of the biosphere if buried. Proof of a hypothesis, via the scientific method, requires demonstration. Yet with radioactive waste, such proof would require hundreds of human generations and entail extensive risks. Critics, from aboriginal people to scientists, have often noted the presumptuousness of our civilization's willingness to reach forward in time, borrowing from the future that which we can never repay. To leave a legacy that does not merely impoverish future life but may endanger it for millennia to come, constitutes an act of unprecedented irresponsibility.

Due to the scientific and political difficulties with geologic burial, above-ground "temporary" storage may remain the only viable option well into the twenty-first century. Yet this option, too, has its dangers. The precariousness of above-ground storage is apparent at weapons facilities in the Soviet Union and the United States. Both governments are under pressure to quickly stabilize liquid waste stored in leaky tanks, and to prevent the further spread of radioactivity from already-contaminated soil and groundwater.

45

Irradiated fuel from civilian nuclear power plants poses a somewhat more manageable problem. All countries operate short-term storage facilities at existing nuclear power plants, though most are susceptible to potentially massive accidents if the cooling system for the thermally hot waste fails. For longer term storage, both governments and independent analysts believe that technologies such as dry casks are safer than water-based systems and are capable of safely containing materials for at least a century, allowing radiation levels to fall 90 percent or more. Even with improved storage systems, however, an institution for the careful monitoring and safeguarding of the waste will be needed to prevent catastrophic accidents or even terrorism. But no government can guarantee the durability of an institution whose responsibilities must continue many times longer than any human institution has ever lasted.[98]

In recent years, a growing number of independent environmental researchers have endorsed the concept of long-term, on-site storage of nuclear waste. Although most countries still officially plan to dismantle nuclear plants shortly after they close, their efforts may be blocked by technical difficulties posed by high levels of radiation found in recently closed reactors, the need to limit worker radiation exposure, the cost of dismantlement, and the lack of facilities to receive radioactive wastes. Not a single large nuclear reactor has been successfully dismantled.[99]

Indeed, old reactors could become permanent fixtures in many countries. The United Kingdom, where commercial nuclear power began in 1956, has become the first nation to give up on the notion of dismantling its reactors and is now planning to entomb them for at least 130 years. Retired reactors could then continue to provide a home to irradiated fuel and other wastes, thereby reducing the number of sites and limiting the handling and transportation of waste. Immediate storage facilities away from existing reactors need only be considered in those cases where reactors are located in seismically active areas or other sites where there is a need for their rapid removal. Short-term storage does not solve the problem of nuclear waste, but it could allow time for more careful consideration of longer-term options, including geologic burial, seabed burial, and indefinite storage.[100]

But addressing the waste problem demands much more than scientific research and reduction of technical uncertainties. It also requires a fundamental change in current operating programs and efforts to regain public confidence. Reordering priorities can only take place once the threat from different types of waste is reevaluated. In the United States, for example, the classification system for wastes needs revamping so that standards are based not on their source, as is done now, but on the actual radioactivities and half-lives of the materials. The existing classification system defies reason by allowing long-lived wastes to be dumped at so-called low-level sites. In Canada, vessel components of the CANDU reactors will be more radioactively dangerous than irradiated fuel 130 years after they are shut down, according to Marvin Resnikoff, a physicist and director of Radioactive Waste Management Associates in New York; yet Canada currently plans on shallow burial of the reactor components upon dismantling.[101]

Regaining public trust will require institutional changes and closer scrutiny of radioactive waste programs. A lack of credibility plagues government nuclear agencies in most countries. Public distrust is rooted in the fact that the institutions in charge also promote nuclear power and weapons production—and have acquired reputations for equivocation, misinformation, and secrecy. The same stigma burdens the IAEA, which could play a far more constructive part in handling radioactive waste, particularly in developing countries, if it no longer had the conflicting roles of both promoter and regulator of nuclear technology.[102]

In the United States, reports by the U.S. Office of Technology Assessment, the National Research Council's Board of Radioactive Waste Management, and private research groups have called for an independent government body to take over the task of managing the country's nuclear wastes. So far, Congress has responded merely by requiring more oversight of the U.S. Department of Energy, while neglecting to tackle the root problem of separating the organization responsible for weapons production and nuclear power promotion from the one that manages waste. Forming autonomous and publicly accountable organizations to handle nuclear waste would go a long way toward regaining public support.[103]

While such organizations would surely be an improvement, questions remain over who would staff them. The independent Nuclear Waste Technical Review Board, authorized by Congress to have 11 members in 1987, had only 8 members in late 1991—partially due to the difficulty of finding qualified board members who do not have a conflict of interest. The number of U.S. university nuclear engineering programs and graduates has fallen by roughly half since the seventies, as the atom has lost its positive image among students, leading the National Research Council to forecast a shortage of nuclear engineers by the mid-nineties. Some European countries face a similar downturn in the number of nuclear graduates.[104]

In the end, the nuclear waste issue is hostage to the overall debate on nuclear power, a debate that increasingly tears at nations. Across Europe and Asia, one country after another has found its political system embroiled over the future of the atom. Because the political controversies are so intense, true progress on the waste issue may only come about once human society turns decisively away from nuclear power. "If industry insists on generating more waste, there will always be confrontation. People just won't accept it," believes British environmental consultant David Lowry. Sweden, which has perhaps the broadest (though not universal) public support for its nuclear waste program, made a national decision to phase out its twelve nuclear power reactors by 2010. Without such a decision, public skepticism toward nuclear technologies and institutions only grows stronger.[105]

Most countries do not have formal policies requiring phase-out of nuclear power. They do, however, have construction programs that are dwindling as rising costs and concern over safety have dried up the supply pipeline. In 1990, for the first time since the dawning of the commercial nuclear age in the mid-1950s, a full year passed without construction starting on a new reactor anywhere in the world. Worldwide, roughly 60 nuclear power plants are under active construction today—fewer than at any other time in more than twenty years.[106]

Despite this trend, nuclear advocates continue to call for a rapid expansion of atomic power. The threat of global warming and public anxieties about dependence on Middle Eastern oil, aroused by the Gulf War, have been seized upon by the nuclear industry as another opportunity to promote itself. Yet a world with six times the current number of reactors, as called for by some nuclear advocates, would require opening a new burial site every two years or so to handle the long-lived wastes generated—a gargantuan financial, environmental, and public health problem that nuclear power proponents discount or even ignore in their calls for revived construction programs. President Bush's 1991 National Energy Strategy, for example, proposed a doubling in the number of U.S. nuclear power plants in the next 40 years, but did not discuss the need for future waste sites. Even if existing burial plans could be carried out as planned, the waste issue would not disappear. As experience with nuclear power plants has demonstrated, it will not necessarily become any easier to site and construct future geologic burial facilities once the first is opened. And a single accident could set back government and industry efforts for decades.[107]

The nuclear waste issue has been marked by a series of illusions and unfulfilled promises. No government has been able to come up with a course of action acceptable either to those dedicated to an expanding nuclear industry or to those determined to stop the production of more nuclear waste. Indeed, a stalemate has formed in nearly every country; attempts to deal with the ever-mounting threat of nuclear waste continue to be pushed further into the future. Without more democratic participation in the development of nuclear waste strategies, publicly acceptable approaches may never be achieved.

Even if no more nuclear waste is created, addressing that which already exists will require attention and investments for a period that defies our usual notion of time. The challenge before human societies is to keep nuclear waste in isolation for the many millennia that make up the hazardous life of these materials. In this light, no matter what becomes of nuclear power, the nuclear age will continue for a very long time. **49**

Notes

50

1. Quote is by Lewis Strauss, commissioner of the U.S. Atomic Energy Commission, before the National Association of Science Writers, New York, September 16, 1954, as cited by Daniel Ford, *The Cult of the Atom* (New York: Simon and Schuster, 1982).

2. Fred C. Shapiro, *Radwaste* (New York: Random House, 1981); Brian Quirke, U.S. Department of Energy, Argonne, Illinois, private communication, November 5, 1991; total irradiated fuel figure of 80,000 tons is Worldwatch Institute estimate based on I.W. Leigh and S.J. Mitchell, Pacific Northwest Laboratory, *International Nuclear Fuel Cycle Fact Book* (Springfield, Va.: National Technical Information Service (NTIS), 1990), on Organisation for Economic Co-operation and Development (OECD), Nuclear Energy Agency (NEA), *Nuclear Spent Fuel Management: Experience and Options* (Paris: 1986), on Andrew Blowers et al., *The International Politics of Nuclear Waste* (New York: St. Martin's Press, 1991), on Soviet figures from G.A. Kaurov, Director of the Center of Public Information for Atomic Energy, Moscow, in letter to Lydia Popova, Socio-Ecological Union, Moscow, August 5, 1991, and on East European production based on United Nations, *Energy Statistics Yearbook* (New York: various years), on British Petroleum (BP), *BP Statistical Review of World Energy* (London: 1991), and on above sources.

3. Matthew L. Wald, "Nature Helps Spread Taint of Nuclear Waste Into the Environment," *New York Times*, December 10, 1988; Zhores A. Medvedev, *The Legacy of Chernobyl* (New York: W.W. Norton, 1990); International Physicians for the Prevention of Nuclear War (IPPNW) and Institute for Energy and Environmental Research (IEER), *Radioactive Heaven and Earth: The Health and Environmental Effects of Nuclear Weapons Testing in, on, and Above the Earth* (New York: Apex Press, 1991).

4. U.S. Department of Energy (DOE), Office of Civilian Radioactive Waste Management (OCRWM), *Integrated Data Base for 1990: U.S. Spent Fuel and Radioactive Waste Inventories, Projections, and Characteristics* (Washington, D.C.: 1990); United Nations, *Energy Statistics Yearbook*; BP, *BP Statistical Review of World Energy*; Ronald L. Fuchs and Kimberly Culbertson-Arendts, *1989 State-by-State Assessment of Low-Level Radioactive Wastes Received at Commercial Disposal Sites* (Idaho Falls, Idaho: DOE, December 1990); estimates for dividing civilian and military waste streams in France, the Soviet Union, and the United Kingdom are impossible due to dual-use nuclear facilities, including reactors and reprocessing plants.

5. Wilson cited in Shapiro, *Radwaste*.

6. For further information on the cost and safety of nuclear power see Christopher Flavin, *Nuclear Power: The Market Test*, Worldwatch Paper 57 (Washington, D.C., December 1983), and Christopher Flavin, *Reassessing Nuclear Power: The Fallout from Chernobyl*, Worldwatch Paper 75 (Washington, D.C.: Worldwatch Institute, March 1987).

7. Nuclear energy contribution is a Worldwatch estimate based on BP, *BP Statistical Review of World Energy*, and on J.M.O. Scurlock and D.O. Hall, "The Contribution of Biomass to

Global Energy Use," *Biomass*, No. 21, 1990; James D. Watkins, Secretary of the Department of Energy, Testimony before the Committee on Energy and Natural Resources, U.S. Senate, Washington, D.C., March 21, 1991; William S. Lee, chairman and president, Duke Power Company, Statement before the Secretary of Energy, National Energy Strategy Hearing, Washington, D.C., August 1, 1989.

8. G. de Marsily et al., "Nuclear Waste Disposal: Can the Geologist Guarantee Isolation?" *Science*, August 5, 1977; Ronnie D. Lipschutz, *Radioactive Waste: Politics, Technology and Risk* (Cambridge, Mass.: Ballinger, 1980).

9. DOE, OCRWM, *Integrated Data Base for 1990*; typical 1,000 megawatt light-water reactor from National Research Council (NRC), Board on Radioactive Waste Management (BRWM), "Rethinking High-Level Radioactive Waste Disposal," National Academy Press, Washington, D.C., July 1990; Lipschutz, *Radioactive Waste*; J.O. Blomeke et al., Oak Ridge National Laboratory, "Projections of Radioactive Wastes to be Generated by the U.S. Nuclear Power Industry," NTIS, Springfield, Va., February 1974; Simon Rippon, "After Five Years, Uncertainties Remain at Chernobyl," *Nuclear News*, June 1991; Christoph Hohenemser, "The Accident at Chernobyl: Health and Environmental Consequences and the Implications for Risk Management," Clark University, Worcester, Mass., December 2, 1987.

10. "World's List of Nuclear Power Plants," *Nuclear News*, August 1991; Leigh and Mitchell, *International Nuclear Fuel Cycle Fact Book*; BP, *BP Statistical Review of World Energy*; DOE, OCRWM, *Integrated Data Base for 1990*; International Atomic Energy Agency (IAEA), *Nuclear Power, Nuclear Fuel Cycle and Waste Management: Status and Trends 1990*, Part C of the *IAEA Yearbook 1990* (Vienna: 1990); Table 2 is based on Leigh and Mitchell, *International Nuclear Fuel Cycle Fact Book*, on OECD, NEA, *Nuclear Spent Fuel Management*, on Blowers et al., *The International Politics of Nuclear Waste*, on Soviet figures from Kaurov, in letter to Popova, and on East European production based on United Nations, *Energy Statistics Yearbook*, on BP, *BP Statistical Review of World Energy*, and on above sources.

11. IAEA, *Nuclear Power, Nuclear Fuel Cycle and Waste Management*; OECD, NEA, *Nuclear Spent Fuel Management*; State of Nevada, Agency for Nuclear Projects/Nuclear Waste Project Office, "Storage of Spent Fuel from the Nation's Nuclear Reactors: Status, Technology, and Policy Options," Carson City, Nev., October 1989.

12. Cumbrians Opposed to a Radioactive Environment (CORE), "A Brief Background to Reprocessing and the Thermal Oxide Reprocessing Plant (THORP)," *Thermal Oxide Reprocessing Plant: An Indepth Investigation* (Cumbria, U.K.: undated); Lipschutz, *Radioactive Waste*.

13. OECD, NEA, *Nuclear Energy in Perspective* (Paris: 1989); Fuchs and Culbertson-Arendts, *1989 State-by-State Assessment of Low-Level Radioactive Wastes*.

14. Fuchs and Culbertson-Arendts, *1989 State-by-State Assessment of Low-Level Radioactive*

Wastes; Arjun Makhijani and Scott Saleska, *High-Level Dollars, Low-Level Sense: A Critique of Present Policy for the Management of Long-Lived Radioactive Wastes and Discussion of an Alternative Approach* (New York: Apex Press, in press); nuclear power's share of low-level waste excludes waste from uranium enrichment for reactor fuel.

15. Edward Landa, *Isolation of Uranium Mill Tailings and Their Component Radionuclides from the Biosphere—Some Earth Science Perspectives*, Geological Survey Circular 814 (Arlington, Va.: United States Geological Service, 1980); Peter Diehl, Herrischried, Germany, private communication and printout, June 28, 1991; "Urals Town Contaminated by Radioactive Waste, to be Evacuated," *Izvestiya*, January 11, 1991, translated in Foreign Broadcast Information Service (FBIS) Daily Report/Soviet Union, Rosslyn, Va., February 5, 1991; DOE, OCRWM, *Integrated Data Base for 1990*; OECD, NEA and IAEA, *Uranium: Resources, Production and Demand* (Paris: 1990).

16. Thomas W. Lippman, "Uranium Pollution Probed at Oklahoma Plant," *Washington Post*, April 29, 1991; Thomas W. Lippman, "NRC Closes Oklahoma Plant After Finding Uranium Leaks," *Washington Post*, October 5, 1991.

17. "World List of Nuclear Power Plants"; IAEA, *Nuclear Power, Nuclear Fuel Cycle and Waste Management*; typical 1,000 megawatt pressurized water reactor is from DOE, OCRWM, *Integrated Data Base for 1990*.

18. U.S. Congress, Office of Technology Assessment (OTA), *Complex Cleanup: The Environmental Legacy of Nuclear Weapons Production* (Washington, D.C.: U.S. Government Printing Office (GPO), 1991); "'Hot Frogs' Loose at Nuclear Laboratory," *Washington Post*, August 4, 1991; Marguerite Holloway, "Hot Geese," *Scientific American*, August 1990.

19. Karen Dorn Steele, "Hanford: America's Nuclear Graveyard," *Bulletin of the Atomic Scientists*, October 1989; Luther J. Carter, *Nuclear Imperatives and Public Trust: Dealing with Radioactive Waste* (Washington, D.C.: Resources for the Future, 1987); DOE, OCRWM, *Integrated Data Base for 1990*; OTA, *Complex Cleanup*; Karen Dorn Steele, "Hanford in Hot Water," *Bulletin of the Atomic Scientists*, May 1991; Scott Saleska and Arjun Makhijani, "Hanford Cleanup: Explosive Solution," *Bulletin of the Atomic Scientists*, October 1990; Keith Schneider, "Military Has New Strategic Goal in Cleanup of Vast Toxic Waste," *New York Times*, August 5, 1991; U.S. General Accounting Office (GAO), "Hanford Single-Shell Tank Leaks Greater than Estimated," Report to the Chairman, Committee on Governmental Affairs, U.S. Senate, August 1991.

20. Thomas B. Cochran and Robert S. Norris, "A First Look at the Soviet Bomb Complex," *Bulletin of the Atomic Scientists*, May 1991; Thomas B. Cochran and Robert Standish Norris, *Nuclear Weapons Databook: Working Papers—Soviet Nuclear Warhead Production*, NWD 90-3, 3rd rev. (Washington, D.C.: Natural Resources Defense Council, 1991); Cochran quote in Matthew L. Wald, "High Radiation Doses Seen for Soviet Arms Workers," *New York Times*, August 16, 1990; Frank P. Falci, "Foreign Trip Report: Travel to USSR for Fact Finding

Discussions on Environmental Restoration and Waste Management, June 15-28, 1990," Office of Technology Development (OTD), DOE, July 27, 1990.

21. Cochran and Norris, *Nuclear Weapons Databook: Working Papers—Soviet Nuclear Warhead Production.*

22. Schneider, "Military Has New Strategic Goal in Cleanup"; OTA, *Complex Cleanup.*

23. Catherine Caufield, *Multiple Exposures: Chronicles of the Radiation Age* (Chicago: University of Chicago Press, 1989).

24. Ibid.; NRC, *Health Effects of Exposure to Low Levels of Ionizing Radiation: BEIR V* (Washington, D.C.: National Academy Press, 1989); Lowell E. Sever, "Low-Level Ionizing Radiation: Paternal Exposure & Children's Health," *Health & Environment Digest*, Freshwater Foundation, Navarre, Minn., February 1991; Alice Stewart, "Low-Level Radiation: The Cancer Controversy," *Bulletin of the Atomic Scientists*, September 1990; Harriet S. Page and Ardyce J. Asire, *Cancer Rates and Risks*, 3rd ed. (Bethesda, Md.: National Institutes of Health (NIH), 1985).

25. Example is based on Scott Saleska et al., "Nuclear Legacy: An Overview of the Places, Problems, and Politics of Radioactive Waste in the U.S.," Public Citizen, Washington, D.C., September 1989; plutonium production is from DOE, OCRWM, *Integrated Data Base for 1990.*

26. "Biological Effects of Radiation," *The New Encyclopedia Britannica*, Macropaedia, Vol. 15 (Chicago: Encyclopedia Britannica, Inc., 1976); Dan Benison, chairman, International Commission on Radiological Protection (ICRP), press conference, Washington, D.C., June 22, 1990.

27. R.H. Clarke and T.R.E. Southwood, "Risks from Ionizing Radiation," *Nature*, March 16, 1989; Stewart, "Low-Level Radiation"; G.W. Kneale and A.M. Stewart, "Childhood Cancers in the U.K. and Their Relation to Background Radiation," Proceedings of the International Conference on Biological Effects of Ionizing Radiation, Hammersmith Hospital, London, November 24-25, 1986.

28. NRC, *BEIR V.*

29. Hylton Smith, Scientific Secretary, ICRP, Didcot, U.K., private communication, March 25, 1991; Caufield, *Multiple Exposures*; John W. Gofman, *Radiation-Induced Cancer from Low-Dose Exposure: An Independent Analysis* (San Francisco: Committee for Nuclear Responsibility, Inc., 1990). One centisievert is equivalent to one rem, another unit used to measure the biological effect of radiation on the body.

30. Steve Wing et al., "Mortality Among Workers at Oak Ridge National Laboratory:

53

Evidence of Radiation Effects in Follow-Up Through 1984," *Journal of the American Medical Association*, March 20, 1991; Thomas W. Lippman, "Risk Found in Low Levels of Radiation," *Washington Post*, March 20, 1991; Peter Aldhous, "Leukemia Cases Linked to Fathers' Radiation Dose," *Nature*, February 22, 1990; Gofman, *Radiation-Induced Cancer*.

31. Maureen C. Hatch et al., "Cancer Rates After the Three Mile Island Nuclear Accident and Proximity of Residence to the Plant," *American Journal of Public Health*, June 1991; Wing et al., "Mortality Among Workers"; Seymour Jablon et al., *Cancer in Populations Living Near Nuclear Facilities* (Bethesda, Md.: NIH, 1990).

32. Robert Peter Gale, "Long-Term Impacts from Chernobyl in U.S.S.R.," *Forum for Applied Research and Public Policy*, Fall 1990; Felicity Barringer, "Chernobyl: The Danger Persists," *New York Times Magazine*, April 14, 1991; Gofman, *Radiation-Induced Cancer*.

33. "Report on Nuclear Program Views Environmental Effects," *O Globo* (Rio de Janeiro), September 30, 1990, translated in FBIS Daily Report/Latin America, Rosslyn, Va., November 2, 1990; Gail Daneker and Jennifer Scarlott, "Nuclear Tragedy Strikes Brazil," *RWC Waste Paper* (Radioactive Waste Campaign, New York), Winter 1987/1988; K.T. Thomas et al., "Radioactive Waste Management in Developing Countries," *IAEA Bulletin*, Vol. 31, No. 4, 1989.

34. Wald, "Nature Helps Spread Taint of Nuclear Waste Into the Environment"; Falci, "Foreign Trip Report: Travel to USSR for Fact Finding Discussions"; estimate for cancer deaths is found in IPPNW and IEER, *Radioactive Heaven and Earth* and is based on risk estimates of *BEIR V*.

35. NRC, BRWM, "Rethinking High-Level Radioactive Waste Disposal"; Table 3 is based on Lipschutz, *Radioactive Waste*, on Frank L. Parker et al., *The Disposal of High-level Radioactive Waste 1984*, Vol. II (Stockholm: Beijer Institute, 1984), on CORE, *Thermal Oxide Reprocessing Plant*, on OECD, NEA, *Feasibility of Disposal of High-Level Radioactive Waste into the Seabed*, Vol. 1 (Paris: 1988), on Frank L. Parker et al., *Technical and Sociopolitical Issues in Radioactive Waste Disposal, 1986*, Vols. IA and II (Stockholm: Beijer Institute, 1987), on IAEA, *Nuclear Power, Nuclear Fuel Cycle and Waste Management*, and on Daniel Gibson, "Can Alchemy Solve the Nuclear Waste Problem?" *Bulletin of Atomic Scientists*, July/August 1991.

36. Table 4 is based on Lipschutz, *Radioactive Waste*, on Report to the Congress by the Secretary of Energy, "Reassessment of the Civilian Radioactive Waste Management Program," Washington, D.C., November 29, 1989, on "World Status of Radioactive Waste Management," *IAEA Bulletin*, Spring 1986, on "World Overview: Radioactive Waste Management," *IAEA News Feature*, May 20, 1988, on Jacques de la Ferté, "What Future for Nuclear Power?" *The OECD Observer*, April/May 1990, on Leigh and Mitchell, *International Nuclear Fuel Cycle Fact Book*, on IAEA, *Nuclear Power, Nuclear Fuel Cycle and Waste Management*, and on OECD, NEA, *Nuclear Spent Fuel Management*.

37. de la Ferté, "What Future for Nuclear Power?"; see also Donald E. Saire, "World Status of Radioactive Waste Management," *IAEA Bulletin*, Spring 1986.

38. Konrad B. Krauskopf, "Disposal of High-level Nuclear Waste: Is It Possible?" *Science*, September 14, 1990.

39. Frank L. Parker et al., *The Disposal of High-Level Radioactive Waste 1984*, Vol. I (Stockholm: Beijer Institute, 1984); Carole Douglis, "Stones that Speak to the Future," *OMNI*, November 1985.

40. Cost estimates are from Makhijani and Saleska, *High-level Dollars; Low-Level Sense*, converted to 1990 dollars.

41. Scott Saleska, Institute for Energy and Environmental Research (IEER), Takoma Park, Md., private communication, July 25, 1991.

42. NRC, BRWM, "Rethinking High-Level Radioactive Waste Disposal."

43. Thomas W. Lippman, "Energy Dept. Set to Ship A-Waste to New Mexico," *Washington Post*, October 4, 1991.

44. Bernd Franke and Arjun Makhijani, "Avoidable Death: A Review of the Selection and Characterization of a Radioactive Waste Repository in West Germany," IEER, Takoma Park, Md., November 1987; Helmut Hirsch, Gruppe Ökologie, Hannover, Germany, private communication, September 25, 1991; Lipschutz, *Radioactive Waste*; Don Hancock, "Getting Rid of the Nuclear Waste Problem: The WIPP Stalemate," *The Workbook*, October/December 1989.

45. Victor S. Rezendes, director, energy issues, GAO, "Nuclear Waste: Delays in Addressing Environmental Requirements and New Safety Concerns Affect DOE's Waste Isolation Pilot Plant," Testimony before the Environment, Energy, and Natural Resources Subcommittee, Committee on Government Operation, U.S. House of Representatives, Washington, D.C., June 13, 1991; Hancock, "Getting Rid of the Nuclear Waste Problem."

46. Eliot Marshall, "The Geopolitics of Nuclear Waste," *Science*, February 22, 1991; Janet Raloff, "Fallout Over Nevada's Nuclear Destiny," *Science News*, January 6, 1990; Archambeau quote in William J. Broad, "A Mountain of Trouble," *New York Times Magazine*, November 18, 1990.

47. Raloff, "Fallout Over Nevada's Nuclear Destiny"; Lipschutz, *Radioactive Waste*.

48. R. Monastersky, "'Young' Volcano Near Nuclear Waste Site," *Science News*, June 30, 1990; de Marsily et al., "Nuclear Waste Disposal"; clairvoyant reference is found in Diane M. Cameron and Barry D. Solomon, "Nuclear Waste Landscapes," in J. Barry

Cullingworth, ed., *Energy, Land, and Public Policy* (New Brunswick, N.J.: Transaction Books, 1990).

49. Lipschutz, *Radioactive Waste*.

50. Ibid.; Alan Riding, "Pact Bans Oil Exploration in Antarctica," *New York Times*, October 5, 1991.

51. Parker et al., *The Disposal of Radioactive Waste 1984*, Vol. II; Lipschutz, *Radioactive Waste*; Thomas W. Lippman, "Disarmament's Fallout," *Washington Post*, October 18, 1991.

52. OECD, NEA, *Feasibility of Disposal of High-Level Radioactive Waste into the Seabed*, Vol. 1; Parker et al., *Technical and Sociopolitical Issues*, Vols. IA and II; J. Kirk Cochran, "Lessons from the Deep Sea," *Nature*, July 19, 1990.

53. Dominique P. Calmet, "Ocean Disposal of Radioactive Waste: Status Report," *IAEA Bulletin*, 4/1989; Katherine Bishop, "U.S. to Determine if Radioactive Waste in Pacific Presents Danger," *New York Times*, January 20, 1991.

54. "London Dumping Convention," *WISE News Communiqué*, December 7, 1990.

55. Denis Hayes, *Nuclear Power: The Fifth Horseman*, Worldwatch Paper 6 (Washington, D.C.: Worldwatch Institute, May 1976); OECD, NEA and IAEA, *Uranium: Resources, Production and Demand*.

56. IAEA, *Nuclear Power, Nuclear Fuel Cycle and Waste Management*; "Nuclear Clean-up Costs May Be $3.4 Billion," *Springville Journal*, Springville, New York, February 29, 1991; Lipschutz, *Radioactive Waste*.

57. Mycle Schneider, WISE, Paris, Letter to Office Parlementaire d'Évaluation des Choix Scintifiques et Technologiques, May 14, 1990; British Central Electricity Generating Board citation is from CORE, *Thermal Oxide Reprocessing Plant*.

58. Colin Sweet, "THORP: A Costly British Error," in CORE, *Thermal Oxide Reprocessing Plant*; "Experts: Direct Disposal Cheaper than Reprocessing," *Nuclear News*, March 1990; Steven Dickman, "Wackersdorf Finally Dies," *Nature*, June 8, 1989.

59. "France May Shut Down Superphenix Permanently," *European Energy Report*, August 24, 1990; "France Plans New Nuclear Waste Law Amid Flurry of Security Reports," *European Energy Report*, February 22 1991; "Clinging to a Forgotten Dream," *Economic and Political Weekly*, December 15, 1990; Cochran and Norris, *Soviet Nuclear Warhead Production*.

60. "Rouvillois Report Predicts Troubles in Coming Years for French Fuel Sector," *Nuclear Fuel*, March 19, 1990; IAEA, *Nuclear Power, Nuclear Fuel Cycle and Waste Management*;

Number of atomic weapons based on 8 kilograms of plutonium needed for each bomb, from Leonard Spector, with Jacqueline R. Smith, *Nuclear Ambitions: The Spread of Nuclear Weapons 1989-1990* (Boulder, Colo: Westview Press, 1990).

61. Lipschutz, *Radioactive Waste*; George W. Rathjens and Marvin M. Miller, "Nuclear Proliferation after the Cold War," *Technology Review*, August/September 1991.

62. Rathjens and Miller, "Nuclear Proliferation after the Cold War"; Spector *Nuclear Ambitions*; R. Jeffrey Smith, "Officials Say Iran Is Seeking Nuclear Weapons Capability," *Washington Post*, October 30, 1991; Jerry Gray, "Baghdad Reveals It Had Plutonium of Weapons Grade," *New York Times*, August 6, 1991; Elaine Sciolino and Eric Schmitt, "Algerian Reactor: A Chinese Export," *New York Times*, November 15, 1991.

63. David Albright, "Civilian Inventories of Plutonium and Highly Enriched Uranium," in Paul Leventhal and Yonah Alexander, eds., *Preventing Nuclear Terrorism* (Lexington, Mass.: Lexington Books, 1987).

64. "French Call for End to Reprocessing and Prices to Reflect Nuclear Cost," *European Energy Report*, March 8, 1991; "UK Nuclear Firms Reduce Losses, Develop Decommissioning Strategy," *European Energy Report*, August 23, 1991; "European Parliament Resolution Calls for an End to all Nuclear Reprocessing," *EC Energy Monthly*, June 1991; Frans Berkhout and William Walker, "Spent Fuel and Plutonium Policies in Western Europe," *Energy Policy*, July/August 1991.

65. "Nuclear to Remain Core of France's Energy Programme, EdF Head Says," *European Energy Report*, April 5, 1991.

66. Parker et al., *The Disposal of Radioactive Waste 1984*, Vol. II; Daniel Gibson, "Can Alchemy Solve the Nuclear Waste Problem?" *Bulletin of Atomic Scientists*, July/August 1991.

67. OECD, NEA, *Nuclear Energy in Perspective*; Gibson, "Can Alchemy Solve the Nuclear Waste Problem?"; Parker et al., *The Disposal of Radioactive Waste 1984*, Vol. II; "Los Alamos Pushes Project for Waste Transmutation," *Energy Daily*, November 1, 1990.

68. Frank L. Parker et al., *Technical and Sociopolitical Issues in Radioactive Waste Disposal, 1986*, Vol. I (Stockholm: Beijer Institute, 1987); Makhijani and Saleska, *High-Level Dollars, Low-Level Sense*; OTA, *Complex Cleanup*.

69. Carter, *Nuclear Imperatives*; National Academy of Sciences report cited in Shapiro, *Radwaste*; 1960 NAS committee cited in Carter, *Nuclear Imperatives*.

70. Cameron and Solomon, "Nuclear Waste Landscapes."

71. DOE, EIA, *Monthly Energy Review September 1991* (Washington, D.C.: 1991); Lipschutz,

Radioactive Waste; Carter, *Nuclear Imperatives*.

72. California law quoted in Shapiro, *Radwaste*; Carter, *Nuclear Imperatives*.

73. See Blowers et al., *The International Politics of Nuclear Waste*, or Gerald Jacob, *Site Unseen: The Politics of Siting a Nuclear Waste Repository* (Pittsburgh: University of Pittsburgh Press, 1990) for a complete review of the 1982 Act.

74. Bob Miller, Governor of Nevada, Testimony before the Committee on Energy and Natural Resources, U.S. Senate, Washington, D.C., March 21, 1991; Paul Slovic et al., "Lessons from Yucca Mountain," *Environment*, April 1991; Victor Gilinsky, "Nuclear Power: What Must Be Done?" *Public Utilities Fortnightly*, June 1, 1991.

75. Duane Chapman, "The Eternity Problem: Nuclear Power Waste Storage," *Comtemporary Policy Issues*, California State University, Long Beach, Calif., July 1990; Jacob, *Site Unseen*.

76. Watkins, Testimony before the Committee on Energy and Natural Resources; Judy England-Joseph, associate director, Energy Issues, Resources, Community, and Economic Development Division, GAO, "Nuclear Waste: DOE Expenditures on the Yucca Mountain Project," Testimony before the Subcommittee on Nuclear Regulation, Committee on Environment and Public Works, U.S. Senate, Washington, D.C., April 18, 1991; Gibson cite in Sandra Sugawara, "Nuclear-Dump Contract Fight Goes to Court," *Washington Post*, March 30, 1989; Paul Rodarte, "Military Maneuvers over Native Lands," *Nuclear Times*, Winter 1990/1991.

77. "Chained to Reactors, *The Economist*, February 2, 1991; "World List of Nuclear Power Plants"; Sylvia Hughes, "Secret Report Attacks French Nuclear Programme," *New Scientist*, March 17, 1990; "Two Reports Call for Reorganisation of French Nuclear Power Industry," *European Energy Report*, January 11, 1991; "French Conservatives Join Greens in Nuclear Waste Disposal Protests," *International Environment Reporter*, November 1989; "Rethink on Waste Storage," *Power in Europe*, February 15, 1990; Ann MacLachlan, "French Government Stops Test Drilling at Waste Sites for 'At Least' a Year," *Nuclear Fuel*, February 19, 1990.

78. "Controversies Surround Nuclear Waste Disposal Plutonium at Saint-Aubin," *Le Monde*, October 20, 1990, translated in FBIS Daily Report/Western Europe, January 4, 1991; "French Waste Problems Continue," *European Energy Report*, January 11, 1991; "French Government Comes Clean over Radioactive Contamination," *New Scientist*, December 1, 1990.

79. Fauroux quote in "Council of Ministers Adopts Draft Law on Disposal of Long-Term Radioactive Waste," *International Environment Reporter*, May 22, 1991; "France Passes New Law on Disposal of Nuclear Waste Underground," *European Energy Report*, July 12, 1991;

William Dawkins, "French to Make Cleaner Job of Nuclear Waste," *Financial Times*, May 15, 1991; Leigh and Mitchell, *International Nuclear Fuel Cycle Fact Book*.

80. H.P. Berg and P. Brennecke, "Planning in-Depth for German Waste Disposal," *Nuclear Engineering International*, March 1991; "Dispute over Nuclear Waste Raging in German State of Lower Saxony," *International Environment Reporter*, July 3, 1991; "East German Nuclear Plant Hopes Fade as Minister Changes Tack," *European Energy Report*, April 19, 1991; Blowers et al., *The International Politics of Nuclear Waste*.

59

81. Berg and Brennecke, "Planning in-Depth for German Waste Disposal"; Hirsch, private communication; "Court Gives Federal Government Go-Ahead on Nuclear Waste Disposal Site," *International Environment Reporter*, April 24, 1991; Parker et al., *The Disposal of Radioactive Waste 1984*, Vol. II; Leigh and Mitchell, *International Nuclear Cycle Fact Book*; "Germany Suspends Storage at East's Morsleben Repository," *Nuclear Europe*, March/April 1991.

82. Carter, *Nuclear Imperatives*; Blowers et al., *The International Politics of Nuclear Waste*; Franke and Makhijani, "Avoidable Death"; Parker et al., *Technical and Sociopolitical Issues*, Vols. IA and II.

83. Carter, *Nuclear Imperatives*; Parker et al., *Technical and Sociopolitical Issues*, Vols. IA and II.

84. NRC, "Review of the Swedish KBS-3 Plan for Final Storage of Spent Nuclear Fuel," NTIS, Springfield, Va., March 1, 1984; National Board for Spent Nuclear Fuel, *Evaluation of SKB R&D Programme 89* (Stockholm: 1990); Blowers et al., *The International Politics of Nuclear Waste*; Parker et al., *Technical and Sociopolitical Issues*, Vols. IA and II.

85. Parker et al., *Technical and Sociopolitical Issues*, Vols. IA and II; Karl-Inge Åhäll et al., *Nuclear Waste in Sweden: The Problem is not Solved!* (Uppsala: The Peoples' Movement Against Nuclear Power and Weapons, 1988); "Swedes Accept Waste, Poll Says," *European Energy Report*, December 14, 1990.

86. "World List of Nuclear Power Plants"; Tatsujiro Suzuki, "Japan's Nuclear Dilemma," *Technology Review*, October 1991; "Utility Says More Nuclear Plants Are Almost Impossible in Japan," *Journal of Commerce*, November 7, 1990; Jacob M. Schlesinger, "Japan Energy Plan Spurs Public Fission," *Wall Street Journal*, January 30, 1991.

87. David Swinbanks, "Yen Melts Down Opposition," *Nature*, May 24, 1990; Schlesinger, "Japan Energy Plan Spurs Public Fission."

88. Swinbanks, "Yen Melts Down Opposition"; Schlesinger, "Japan Energy Plan Spurs Public Fission"; Yoji Takemoto, Wisetokyo, Tokyo, private communication, February 14, 1991.

89. "HLW Disposal Plans Come to Light," *Nuke Info Tokyo*, November/December 1989;

"Hokkaido Government Opposes HLW Plan in Horonobe," *Nuke Info Tokyo*, September/October 1990.

60

90. "German Spent Fuel May Go to China," *Nuclear News*, October 1985; "Joint Research Agreement Reached with China on HLW Disposal," *Nuke Info Tokyo*, November/December 1990.

91. "World List of Nuclear Power Plants"; Charles Mitchell, "Fallout From Chernobyl Accident Still Clouds Soviet Nuclear Plans," *Journal of Commerce*, April 19, 1991; Gabriel Schonfeld, "Rad Storm Rising," *Atlantic*, December 1990; "Urals Town Contaminated by Radioactive Waste, to be Evacuated."

92. Cochran and Norris, *Nuclear Weapons Databook: Working Papers—Soviet Nuclear Warhead Production*; "Resolution on Nuclear Waste State Program Viewed," *Sovestskaya Rossiya*, June 28, 1990, translated in FBIS Daily Report/Regional Affairs, Rosslyn, Va., July 3, 1990; "Soviet Plans to Construct Waste Facility in Siberia Protested," *Multinational Environmental Outlook*, July 11, 1989; "Oslo Concern on USSR Waste Storage," *European Energy Report*, Eastern Europe Supplement, December 14, 1990; "Nuclear Plans Divide Baltic Neighbors," *New Scientist*, December 1, 1990.

93. Michael Wise, "Nuclear Waste Piles Up in Eastern Europe," *Washington Post*, July 17, 1991; "Resolution on Nuclear Waste State Program Viewed"; "Radioactive Waste Found in Cesky Kras," *Zemedelske Noviny*, Prague, December 4, 1990, translated in FBIS Daily Report/East Europe, Rosslyn, Va., February 13, 1991; "Pollution Found at Bulgarian N-Plant," *Financial Times*, July 25, 1991.

94. Mark Clifford, "A Nuclear Falling Out," *Far Eastern Economic Review*, May 18, 1989; "Nuclear Waste Policy Examined in Light of Protest," *The Korea Times*, November 10, 1990, in FBIS Daily Report/East Asia, Rosslyn, Va., November 21, 1990; Huh Sook, "Overview of Korea Nuclear Program," *Nuclear Europe*, March/April 1991; "Nuclear Dumps to be Built on Uninhabited Islands," *Korea Herald*, November 30, 1990, in FBIS Daily Report/East Asia, Rosslyn, Va., January 4, 1991.

95. "Argentine Activist Receives Death Threats," *WISE News Communiqué*, December 21, 1990; C.S. Lee, "Taipower's Backend Management," *Nuclear Europe*, March/April 1991; Carl Goldstein, "Nuclear Qualms," *Far Eastern Economic Review*, July 4, 1991; "Lanyu Reps Demand Taipower Remove Nuclear Dump," *Occasional Bulletin*, Taiwan Church News, Tainan, September, October 1991; Chris Brown, "Deadly Anti-Nuclear Protest Further Stalls Taiwan Plant," *Journal of Commerce*, October 4, 1991.

96. "World List of Nuclear Power Plants"; Prakash Chandra, "India: Going Nuclear," *Third World Week*, December 23, 1990; Barbara Crossette, "300 Factories Add Up to India's Very Sick Town," *New York Times*, February 6, 1991; Barry D. Solomon and Fred M. Shelley, "Siting Patterns of Nuclear Waste Repositories," *Journal of Geography*, March-April 1988;

Department of Atomic Energy, *Annual Report: 1989-1990* (New Delhi, 1990).

97. Chenevier quoted in William Dawkins, "Loud Rumblings from Beneath the Surface," *Financial Times,* December 5, 1990.

98. Makhijani and Saleska, *High-Level Dollars, Low-Level Sense;* Parker et al., *Technical and Sociopolitical Issues,* Vol. I.

99. Makhijani and Saleska, *High-Level Dollars, Low-Level Sense;* Greenpeace and Friends of the Earth, "Radioactive Waste Management: The Environmental Approach," London, November 1987; Gordon MacKerron, "Decommissioning Costs and British Nuclear Policy," *The Energy Journal,* Vol. 12, Special Issue, 1991; DOE, EIA *Commercial Nuclear Power* (Washington, D.C.: 1990); for more information on decommissioning, see Martin J. Pasqualetti and Geoffrey S. Rothwell, eds., *The Energy Journal,* Vol. 12, Special Issue, 1991, or Cynthia Pollack, *Decommissioning: Nuclear Power's Missing Link,* Worldwatch Paper 69 (Washington, D.C.: Worldwatch Institute, April 1986).

100. "NE Wants Decommissioning to be Slower, Cheaper," *Nuclear News,* August 1991; Andrew Holmes, "Take It Away, Kids!" *Energy Economist,* July 1991; Makhijani and Saleska, *High-Level Dollars, Low-Level Sense.*

101. Makhijani and Saleska, *High-Level Dollars, Low-Level Sense;* Marvin Resnikoff, Radioactive Waste Management Associates, "Memorandum: CANDU Decommissioning," New York, April 22, 1991.

102. Blowers et al., *The International Politics of Nuclear Waste;* "French Ministry of Industry to Prepare National Nuclear Report," *European Energy Report,* November 16, 1990; "Two Reports Call for Reorganisation of French Nuclear Power Industry"; David E. Sanger, "A Crack in Japan's Nuclear Sangfroid," *New York Times,* February 17, 1991; Robert Thomson, "Nuclear Mishap Shakes Confidence in Japan's Programme," *Financial Times Weekend,* February 16-17, 1991; "Travel to the USSR for the First Fact Finding Meeting of the US-USSR Joint Coordinating Committee on Environmental Restoration and Waste Management, November 8-17, 1990," OTD, DOE, December 17, 1990; Medvedev, *The Legacy of Chernobyl;* Radioactive Waste Campaign, *Deadly Defense: Military Radioactive Landfills* (New York: 1988).

103. See, for instance, NRC, BRWM, "Rethinking High-Level Waste", OTA, *Managing the Nation's Commercial High-Level Radioactive Waste* (Washington, D.C.: GPO, 1985), OTA, *Complex Cleanup,* Arjun Makhijani, "Reducing the Risks: Policies for the Management of Highly Radioactive Nuclear Waste," IEER, Takoma Park, Md., May 1989, Carter, *Nuclear Imperatives,* and Jacob, *Site Unseen,* which includes reference to OTA testimony before the U.S. Congress in 1981.

104. Jacob, *Site Unseen;* Joanne Donnelly, National Waste Technical Review Board,

Arlington, Va., private communication, August 1, 1991; Ann Pomeroy, "Nuclear Engineer Shortage Likely by Mid-1990s," *NewsReport*, National Research Council, Washington, D.C., October 1990; Peter Palinkas, Directorate General for Research, European Parliament, Luxembourg, private communication, September 10, 1991; Uwe Fritsche, Öko Institut, Darmstadt, Germany, private communication, September 11, 1991.

105. Blowers et al., *The International Politics of Nuclear Waste*; Mitchell, "Fallout From Chernobyl Accident Still Clouds Soviet Nuclear Plans"; "Ukraine Halts Nuclear Programme as Energy Sovereignty Grows," *European Energy Report*, September 21, 1990; David Lowry, environmental consultant, Milton Keyes, U.K., private communication, Washington, D.C., April 29, 1991.

106. Number of nuclear power plants under construction is a Worldwatch Institute estimate based on "World List of Nuclear Power Plants" and other sources; construction starts is from IAEA, *Nuclear Power Reactors in the World*, April 1991.

107. Jacob, *Site Unseen*; Wolf Häfele, "Energy from Nuclear Power," *Scientific American*, September 1990; DOE, *National Energy Strategy: Powerful Ideas for America*, First Edition 1991/1992 (Washington, D.C.: GPO, 1991).

NICHOLAS LENSSEN is a Research Associate with the Worldwatch Institute, and coauthor of Worldwatch Paper 100, *Beyond the Petroleum Age: Designing a Solar Economy*. He is a graduate of Dartmouth College, where he received a degree in geography.

THE WORLDWATCH PAPER SERIES

No. of
Copies

_____ 57. **Nuclear Power: The Market Test** by Christopher Flavin.

_____ 58. **Air Pollution, Acid Rain, and the Future of Forests** by Sandra Postel.

_____ 60. **Soil Erosion: Quiet Crisis in the World Economy** by Lester R. Brown and Edward C. Wolf.

_____ 61. **Electricity's Future: The Shift to Efficiency and Small-Scale Power** by Christopher Flavin.

_____ 62. **Water: Rethinking Management in an Age of Scarcity** by Sandra Postel.

_____ 63. **Energy Productivity: Key to Environmental Protection and Economic Progress** by William U. Chandler.

_____ 65. **Reversing Africa's Decline** by Lester R. Brown and Edward C. Wolf.

_____ 66. **World Oil: Coping With the Dangers of Success** by Christopher Flavin.

_____ 67. **Conserving Water: The Untapped Alternative** by Sandra Postel.

_____ 68. **Banishing Tobacco** by William U. Chandler.

_____ 69. **Decommissioning: Nuclear Power's Missing Link** by Cynthia Pollock.

_____ 70. **Electricity For A Developing World: New Directions** by Christopher Flavin.

_____ 71. **Altering the Earth's Chemistry: Assessing the Risks** by Sandra Postel.

_____ 73. **Beyond the Green Revolution: New Approaches for Third World Agriculture** by Edward C. Wolf.

_____ 74. **Our Demographically Divided World** by Lester R. Brown and Jodi L. Jacobson.

_____ 75. **Reassessing Nuclear Power: The Fallout From Chernobyl** by Christopher Flavin.

_____ 76. **Mining Urban Wastes: The Potential for Recycling** by Cynthia Pollock.

_____ 77. **The Future of Urbanization: Facing the Ecological and Economic Constraints** by Lester R. Brown and Jodi L. Jacobson.

_____ 78. **On the Brink of Extinction: Conserving The Diversity of Life** by Edward C. Wolf.

_____ 79. **Defusing the Toxics Threat: Controlling Pesticides and Industrial Waste** by Sandra Postel.

_____ 80. **Planning the Global Family** by Jodi L. Jacobson.

_____ 81. **Renewable Energy: Today's Contribution, Tomorrow's Promise** by Cynthia Pollock Shea.

_____ 82. **Building on Success: The Age of Energy Efficiency** by Christopher Flavin and Alan B. Durning.

_____ 83. **Reforesting the Earth** by Sandra Postel and Lori Heise.

_____ 84. **Rethinking the Role of the Automobile** by Michael Renner.

_____ 85. **The Changing World Food Prospect: The Nineties and Beyond** by Lester R. Brown.

_____ 86. **Environmental Refugees: A Yardstick of Habitability** by Jodi L. Jacobson.

_____ 87. **Protecting Life on Earth: Steps to Save the Ozone Layer** by Cynthia Pollock Shea.

_____ 88. **Action at the Grassroots: Fighting Poverty and Environmental Decline** by Alan B. Durning.

_____ 89. **National Security: The Economic and Environmental Dimensions** by Michael Renner.

_____ 90. **The Bicycle: Vehicle for a Small Planet** by Marcia D. Lowe.

_____ 91. **Slowing Global Warming: A Worldwide Strategy** by Christopher Flavin.
_____ 92. **Poverty and the Environment: Reversing the Downward Spiral** by
        Alan B. Durning.
_____ 93. **Water for Agriculture: Facing the Limits** by Sandra Postel.
_____ 94. **Clearing the Air: A Global Agenda** by Hilary F. French.
_____ 95. **Apartheid's Environmental Toll** by Alan B. Durning.
_____ 96. **Swords Into Plowshares: Converting to a Peace Economy** by
        Michael Renner.
_____ 97. **The Global Politics of Abortion** by Jodi L. Jacobson.
_____ 98. **Alternatives to the Automobile: Transport for Livable Cities** by
        Marcia D. Lowe.
_____ 99. **Green Revolutions: Environmental Reconstruction in Eastern
        Europe and the Soviet Union** by Hilary F. French.
_____100. **Beyond the Petroleum Age: Designing a Solar Economy** by
        Christopher Flavin and Nicholas Lenssen.
_____101. **Discarding the Throwaway Society** by John E. Young.
_____102. **Women's Reproductive Health: The Silent Emergency** by Jodi L. Jacobson.
_____103. **Taking Stock: Animal Farming and the Environment** by Alan B. Durning and .
        Holly B. Brough
_____104. **Jobs in a Sustainable Economy** by Michael Renner
_____105. **Shaping Cities: The Environmental and Human Dimensions** by Marcia D. Lowe
_____106. **Nuclear Waste: The Problem That Won't Go Away** by Nicholas Lenssen

_____ **Total Copies**

☐ **Single Copy: $5.00**
☐ **Bulk Copies (any combination of titles)**
    ☐ 2–5: $4.00 each   ☐ 6–20: $3.00 each   ☐ 21 or more: $2.00 each

☐ **Membership in the Worldwatch Library: $25.00 (overseas airmail $40.00)**
  The paperback edition of our 250- page "annual physical of the planet,"
  State of the World 1991, plus all Worldwatch Papers released during
  the calendar year.

☐ **Subscription to *World Watch* Magazine: $15.00 (overseas airmail $30.00)**
  Stay abreast of global environmental trends and issues with our award-winning,
  eminently readable bimonthly magazine.

No postage required on prepaid orders. Minimum $3 postage and handling
charge on unpaid orders.

Make check payable to Worldwatch Institute
1776 Massachusetts Avenue, N.W., Washington, D.C. 20036-1904 USA

Enclosed is my check for U.S. $_____

_____

name   **daytime phone #**

_____

address

_____

city   **state**   **zip/country**